Appalachian
Spring

Appalachian Spring

MARCIA BONTA

University of Pittsburgh Press

Published by the University of Pittsburgh Press, Pittsburgh, Pa. 15260

Copyright © 1991, Marcia Bonta

All rights reserved

Baker & Taylor International, London

Manufactured in the United States of America

LIBRARY OF CONGRESS CATALOGING-IN-PUBLICATION DATA

Bonta, Marcia, 1940–
    Appalachian Spring / Marcia Bonta.
        p.   cm.
Includes bibliographical references and index.
I S B N 0-8229-3658-5.— I S B N 0-8229-5442-7 (paper)
1. Natural history—Appalachian Region.   2. Forest ecology—
Appalachian Region.   3. Spring—Appalachian Region.   I. Title.
QH104.5.A6B66   1991
508.748—dc20   90-12478
                                                                                    CIP

Portions of this book have appeared in a slightly different form in the *Altoona Mirror, Bird Watcher's Digest, Pennsylvania Birds, Pennsylvania Game News,* and *WildBird.*

To my husband Bruce
and our three sons, Steven, David, and Mark.
Without their perceptions
my springs would have had less meaning.

Contents

Acknowledgments

EVERYONE who writes books has a special favorite and, from the moment I conceived the idea of writing about a typical Appalachian mountain spring, I thought of this book as the book of my heart. I especially thank my husband Bruce, who encouraged me to go ahead with it and, as usual, read it carefully and critically. So did my son David. Each, in their own way, saw things that I did not. The book is the better for both their editorial eyes.

Catherine Marshall, managing editor at the University of Pittsburgh Press, is the best book editor I have ever worked with. She questioned the obscure and cleaned up my prose when it needed it, but she never altered my voice.

Magazine editors who have encouraged me in my nature writing include Bob Bell and Bob Mitchell of *Pennsylvania Game News,* Mary Beacom Bowers of *Bird Watcher's Digest,* Bob Carpenter of *WildBird,* and Barbara M. Haas of *Pennsylvania Birds*.

I would also like to thank the many loyal readers of the column I wrote for the *Altoona Mirror* from 1978 until 1986. In many ways, this book is an outgrowth of that effort and my attempt to give those readers a permanent record of at least some of the creatures and plants I wrote about over the years.

My son Mark, who is a geographer, drafted the map of our property. I was fortunate to have a mapmaker intimately acquainted with the mountain and I am grateful for his accurateness and expertise. Bill Nelson prepared the final version.

Most of all, I thank Frederick A. Hetzel, director of the University of Pittsburgh Press, for his understanding and support. He, like me, is a Pennsylvaniaphile and dedicated to publishing books by and for Pennsylvanians that celebrate the beauty of the Commonwealth.

Introduction

TWENTY YEARS AGO, on a Fourth of July weekend, we—my husband, Bruce, our three small sons, and I—discovered our mountaintop home-to-be. Desperate to live in the country, I had called dozens of realtors within a twenty-five mile radius of Bruce's new job in central Pennsylvania and had learned of a very isolated house in the woods that no one was interested in.

I assured the dubious realtor we *were* interested, and I managed to cajole directions to the place from him. Thus we found ourselves creeping up a narrow gravel road cut into the side of a steep mountain slope which followed the meandering of a small stream. Vegetation pressed in on either side, and we drove for what seemed like miles through a dark tunnel of foliage.

At last we saw sunlight flooding a large meadow ahead of us. We emerged into a clearing dominated by a white clapboard house set on a knoll, looking, with its pillared veranda, like a miniature Southern mansion. Below it stood a whitewashed stone springhouse. A smaller, older house with board-and-batten siding and six-pane sash windows was the first of three buildings on the opposite side of the driveway. Directly above that house was a long narrow lawn leading to a shed with a pointed roof topped by a cupola. A matching cupola adorned the much larger bank barn across the barnyard from the shed.

The impact of that first look immediately convinced all of us that we had to buy the place. By the following week we were the proud owners of the old Plummer estate which included 162 acres, more or less, of woods and fields. Later we discovered a variety of mini-habitats to explore along miles of well-kept trails that crisscrossed the mountaintop.

Each of us formed our own attachment to the place. Our eldest son, Steve, continued his passionate interests in bird-watching and insect-collecting. David, our second son, nurtured his poetic nature through a lively appreciation of the beauty around him, and Mark, the classifier, found his niche by keeping lists of every plant and wild creature living on the mountain.

My husband became a champion of trees and specifically of the lush, second-growth forest on either side of our mile-and-a-half access road up what is called Plummers Hollow: old hemlocks and beeches, white oaks, and tulip poplars. When the beauty of the hollow was threatened by absentee land-owners who wanted to lumber its steep slopes, first he fought them to a standstill and then purchased 350 more acres of mountain land to save it from the chain saw.

But of all the family members, I was the one most obsessed with the beauty that surrounded us. I was out in all seasons of the year, compulsively recording what I observed of the natural life of an Appalachian mountain, first in a journal, then in local newspaper columns and magazine articles, and finally in my first book. Without, in fact, ever planning to be one, I found myself a writer—a naturalist-writer, I always hasten to tell people whenever they ask. To me the nature comes first, the writing second. Writing is merely a tool I use to tell others about the wonders of the natural world.

It seems to me that no wild areas can be permanently saved unless people appreciate and understand a little about nature for its own sake and not for what they can take from it. For centuries humanity existed in what I call the first stage, exploiting the natural world beyond its limits to reproduce. Then most, but not all of us, moved into the second stage where the buzz word became "harvest." To harvest properly, land and its wildlife had to be managed, and since only "harvestable" crops—trees, game animals, and fish—were considered, the survival of "non-useful" plants and creatures was incidental.

Today a few of us are moving slowly into the third stage, that of empathy with all of nature for its own sake. Although we call ourselves "environmentalists," the less sophisticated and often derisive title of "nature-lover" is probably more suitable. What, after all, has humanity held as the highest good but love, pure and unselfish—the kind of love that most religions on earth urge us to feel toward our fellow human beings.

We "nature-lovers" take it a step farther. Love the earth, we exhort. To know the earth better, to grasp a little of its workings, to look on it with awe and wonder as well as with respect is to want to save it from destruction. Yet our numbers are still minuscule in comparison with all those who would milk the earth for their own profit. "I see mature trees," one lumberman told us, "as a field of corn, put here by God for man to harvest." To him a managed world is a beautiful world.

There is no doubt that my own piece of the Appalachians has been and will continue to be managed in part by us and by the laws of our state. My husband cuts our large First Field on a four-year rotation, managing not for domestic crops but for wild plants and animals. He also keeps our paths reasonably mowed, our road reasonably clear of debris, and our Far Field half in meadow flowers, half in locust trees. We do not post our land against hunters, not because we believe in hunting but because we know that our state overmanages the deer herd. If I had my way we would bring back the natural predators, but that is no longer possible.

What *is* possible is the further saving from destruction of pockets of wild land by people who watch rather than manage the land, who make no demands other than the right to have such unmanaged land available to them. For those kinds of people, as well as for those who are still in the second and first stages, I have written my own love song about the place and season on earth I love the most.

Appalachian Spring

Prelude

The advent of a North American spring is still the greatest show on earth, a show so compelling that I never turn it off, a show whose preliminaries begin, in fits and starts, soon after New Year's Day on the first clear day in January. As I stand at the top of First Field, bundled against the cold, I watch the light flood across the landscape, setting the field ablaze and casting long blue shadows in the woods, and I realize that the worst of winter—its early, dreary dusks and late, gray sunrises—has passed with December's passing. Already the upward spiral toward spring has begun.

Early in January I hear the first intimation of spring. Great horned owls are courting on Sapsucker Ridge, their calls echoing down across the silent, moon-drenched, winter fields. First come the alto tones of the larger female, then the answering bass notes of the male. Those penetrating "whoo! whoo! whoo-whoo!" calls have tremendous carrying power, and I hear them through the walls of my winter-tight home. I rush outside to listen, summoned by the first pulsating reminders of imminent spring.

No matter the snow, wind, and cold of January and February, the great horned owls will go about their appointed business of courtship, mating, and egg-laying. They mate for life, and both parents raise the young. By mid-February, in our state, the female has laid her two to three white eggs in the abandoned nest of a red-tailed hawk, crow, or squirrel high in a tree. She broods them while the male keeps her fed and occasionally relieves her on the nest. The eggs are incubated for a little over a month, and the young spend another six to seven weeks in the nest. By the time they fledge, spring has truly arrived, but the leaves have not yet emerged. This makes hunting prey easier for the inexperienced youngsters.

The mountain ridges here have what wildlife managers call second-choice habitat for great horned owls—small white oaks and beeches. Because these trees retain many of their leaves throughout the winter, they provide excellent camouflage for resting owls during the day. In the hollow, where they also live, the habitat is even better because it contains two of their favorite nesting and perching trees—white pines and hemlocks. Studies show that their favorite prey animals are white-footed mice, meadow voles, and cottontail rabbits, all of which are abundant here. They also prey on the smaller barred and screech owls, as well as mammals such as porcupines, skunks, squirrels, weasels, woodchucks, rats, and shrews. In fact, great horned owls are opportunists where food is concerned, and one wintry dusk I watched an owl struggle to lift one of my Muscovy ducks which fought so valiantly that the "feathered tiger," as it is sometimes called, gave up and flew off.

Because great horned owls are at the top of the predator chain on the mountain, they have no natural enemies. Up until 1965 they were fair game at any time of the year, but then Pennsylvania discontinued its bounty on what was perceived (and still is by many) as a menace to the game wildlife population. Yet they continue to be shot despite federal laws against shooting any birds except game birds in season or Eu-

ropean starlings and house sparrows. They are also susceptible to poisons, which they ingest through the birds and rodents they eat, which in turn have eaten poisoned insects or
bait. Automobiles, trapping, and flying into telephone and
electrical wires are other hazards. The two dead great horned
owls I have found here had their necks broken by the electric
line that stretches from the garage to the barn.

Despite those two deaths, the mountain is never without
its great horned owl population. Any empty niche is quickly
filled by young birds that must find their own territory and
mate. Those calls I listen to also establish territory, help to
renew pair bonds, and momentarily frighten their prey, making them more vulnerable to capture. Without the voice of
great horned owls to break the winter silences and remind me
of the imminence of spring, my life here would be greatly
diminished.

Early one morning in mid-January as I stand on the veranda, a certain feeling in the air tells me that this day will be
warm—a January thaw is in the offing.

Our home sits tucked into a small hollow between two
ridges. The sun rises behind Laurel Ridge, and before we can
glimpse it from the house its first rays burnish the treetops of
Sapsucker Ridge with a reddish glow. During a January thaw
the dawn light is scarlet as it moves down the ridge and across
First Field, and I go out to meet it. I sit with my back against
a locust tree at the edge of the field, bathed in warm sunlight,
while the house below me is still in shadow. In patches where
the snow has melted, the exposed dried grasses are covered
with frost crystals shining in the light and adding glitter to an
already glorious scene.

Except for the springlike sound of an occasional "peter-
peter" from a tufted titmouse down near the house, I have a
silent vigil beneath the locust tree, memorizing as best I can
the warm touch of sunshine and the constantly shifting, glow-

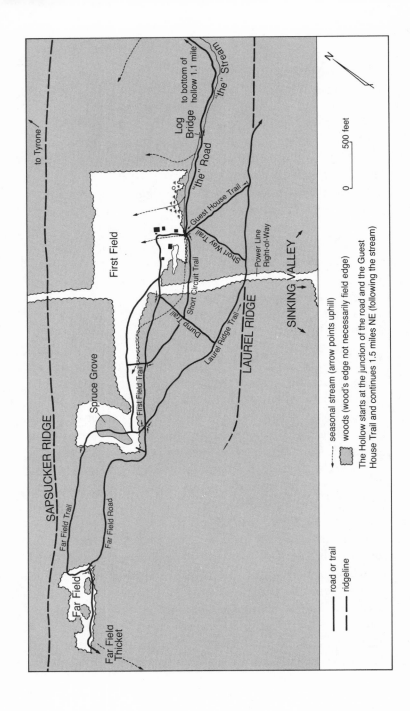

SAPSUCKER RIDGE

Far Field Trail

Far Field Road

Far Field

Far Field Thicket

Spruce Grove

First Field Trail

First Field

to Tyrone

Log Bridge

to bottom of hollow 1.1 mile

"the" Stream

"the" Road

Guest House Trail

Short Way Trail

Short Circuit Trail

Dump Trail

Laurel Ridge Trail

Power Line Right-of-Way

LAUREL RIDGE

SINKING VALLEY

N

——— road or trail

– – – ridgeline

········· seasonal stream (arrow points uphill)

[shaded] woods (wood's edge not necessarily field edge)

The Hollow starts at the junction of the road and the Guest House Trail and continues 1.5 miles NE (following the stream)

0   500 feet

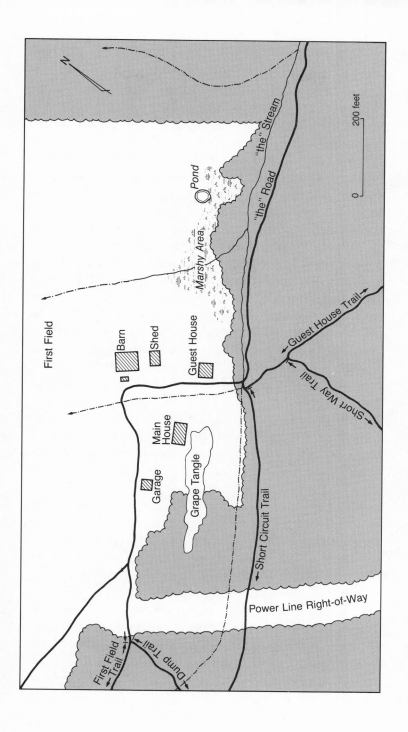

ing light around me. When at last the sun's rays reach the house, I rise and follow it homeward, feeling I have begun the day in the best possible way.

The thermometer rises rapidly in the warmth and by mid-morning it is fifty degrees. Once again I pull on my boots and head outside, eager to absorb as much sunshine as possible. It looks like spring, with the ground thawing and the red maple buds a swollen glow against the ridges. The view from the mountaintop is shrouded by a springlike haze. It smells like spring, moist and muddy, and it certainly feels like spring.

But most of all it sounds like spring. The lone wintering song sparrow in the grape tangle sings its spring carol over and over, and pileated woodpeckers in the woods keep up a rapid-fire drumming unlike the usual random hammering noises they make in winter when searching the trees for car-penter ants. Instead of one tufted titmouse giving its call, the woods echo with titmice song. The black-capped chickadees change their "dee-dee" calls to their clear "fee-bee" song. My heart rejoices, "It's spring"; but my head replies, "Only Jan-uary."

Only January, indeed, and only a January thaw to lift my sometimes winter-weary spirits. But it is a promise of more to come, and I fall asleep that night with the sound of flowing water in my ears. Spring music!

Much of what goes on in the animal world is hidden from me; but in winter, animal tracks in the snow make it possible for me to know when the red foxes are courting or the male woodchucks are wandering in search of females. Several times, though, in late January, I have also been given a ring-side seat to gray squirrel courtship.

I remember one such time particularly well—a thirty-degree, clear, and beautiful morning with two to three inches of snow on the ground when I set out on my usual three-mile, late morning walk over the mountaintop trails. The woods were silent except for a mixed flock of brown creepers, white-breasted nuthatches, and black-capped chickadees that I encountered where Laurel Ridge Trail merges into the Far Field Road. The latter is a flat jeep track winding above a wooded hollow which offers me, especially in winter, a clear view below of grapevine-entangled oak woods and fallen trees. The wild creatures seem to love this road as much as I do, and it is always riddled with the tracks of white-tailed deer and wild turkeys, red foxes and gray squirrels. Often I see those animals on the trail ahead of me, but I never hear them. So I was surprised when the silence was broken by unknown animal noises in the woods below the road.

Immediately I sat down on a fallen log facing the sound and traced it to three gray squirrels frolicking in a large tree. Gray squirrels, according to Joseph F. Merritt, in his *Guide to the Mammals of Pennsylvania,* are not as noisy as red squirrels, but they do have alarm calls which include barks, grunts, and possibly even a song. They also talk among themselves in light chuckles that sound almost musical, as described by Doutt, Heppenstall, and Guilday in their *Mammals of Pennsylvania.* But the sounds I heard could only be described as begging noises, and as I watched I decided I was observing courtship rites.

Gray squirrels are usually solitary creatures, but during mating times (here on the mountain in January and February and again in June and July) several males will chase a female in heat. In this case, it looked as if two males were vying for the attention of one female. For several minutes they seemed to be chasing for the sheer fun and exercise of it. But then one squirrel suddenly paused on a branch face-to-face with a second squirrel and slowly flicked its tail like an undulating wave. The third squirrel scampered on down the tree trunk and ran off into the woods, leaving the two remaining squirrels to continue the chase.

With the disappearance of that squirrel, the chasing slowed

down. The longer I watched, the more the chasing resembled a stylized dance, like ballet, with every motion known ahead of time by the participants. Not only was the chasing slow and graceful but it was punctuated with long pauses, when one squirrel (the female, I presumed) would lie out along a branch with her tail flat over her back and head while the other squirrel would climb to a branch directly above and peer down at her. The first pause lasted five minutes and was broken when the prone squirrel climbed up to face the seated squirrel for a silent second or two, then clambered several yards back down the tree to another, much lower branch.

Before it settled into the same prone, tail-over-body-and-head position, it made a noise like a plunked banjo string, and the other squirrel sat up more alertly. Silence and stillness for close to ten minutes prevailed before the stylized, slow-motion chasing resumed, this time accompanied by whimpering noises. As they chased, their bodies rippled underneath as well as on top of the branches, reminding me of furry serpents.

For the third time they stopped, returned to their respective branches and positions and an even longer period of stillness. And then the scene was abruptly ended when the third squirrel suddenly streaked back up the tree. All three squirrels ran to the ground, squeaking like overgrown mice, and disappeared into the brushwood.

As soon as I returned home, I searched the literature for accounts of gray squirrel courtship behavior and learned only that gray squirrels always disappear when they see people; consequently, no human has ever witnessed gray squirrel courtship. But I had been sitting out in the open, had shifted around and even moved to a more comfortable log, and had had the distinct impression that the squirrels had known I was there and had not cared. I was disappointed that I had seen no actual mating so I could not be positive about what had been going on, but the books did say that gray squirrels always mated in the privacy of their nests. All I can do is assume what seemed obvious to me and to hope that someday my observations will be duplicated by a bona fide biologist. Apparently to have shared such moments with gray squirrels is a privilege granted to very few people.

Great horned owls and gray squirrels are not the only creatures preoccupied by procreation in January. One night a foot of snow falls and once it clears the wind still howls, forcing me to walk down the hollow road which is reasonably sheltered. Carefully I poke along in the fresh tire tracks my husband made earlier on his way to work. I quickly discover that the tracks are being utilized as a pathway by winter crane flies or snow flies *(Chionea valga)* which, on cursory inspection, look more like spiders than flies. According to Ann Haven Morgan's seminal book on animals in winter, these flies often walk over the snow during mild winter days, having crept upward from hiding places at the bases of tree trunks as the temperature rises during snowstorms. Once the sun shines, the light entices them to crawl over the snow in search of mates. After copulation, the females bury beneath the snow to deposit their eggs close to tree trunks.

By February, spring is definitely on the move although we have robins with us every winter so I no longer consider their presence a reliable harbinger of the season. But when I hear them singing, as they do in early February, I know that spring is coming on. Add to that the first cardinals in song as they eat the last of the wild, bittersweet berries growing at the upper edge of First Field and I feel more and more certainty that winter's back is broken. Many of our wintering birds, like me, react more to the lengthening light than they do to the still wintry temperature.

Then one afternoon, as I sit reading *American Entomologists* beside the warmth of the woodstove, a ladybug, more correctly called a ladybird beetle, crawls across the page. Knowing of this creature's very unladylike ferociousness toward the aphid world, I tenderly scoop it up and deposit it on my aphid-infested Christmas cactus. Among our most common insects, ladybird beetles are members of the family Coccinellidae which, appropriately enough, is Greek for "scarlet," the color of most of the species—including the one that interrupts my reading. Apparently encouraged by the warmth in the room, this beetle has crept out from hibernation and had coincidentally landed on my book about insect lovers.

Mid-February, and I am out every day watching for red fox sign. During my first fifteen years here, I had only a passing acquaintanceship with red foxes—a glimpse of one hunting in First Field one early spring morning, the sight of crows mobbing another on the Far Field Trail, a face-to-face meeting that startled both the fox and me on Laurel Ridge Trail—nothing, in fact, but the frustratingly brief encounters that humans so often have with wild animals.

For several years, though, I had been aware of what looked like the remnants of an old fox den above an isolated, overgrown field that we call the Far Field. The den faces south and is several hundred feet uphill from the beginnings of a mountain stream. So I kept a careful watch on it, looking for fox sign, and after years of hoping, I was rewarded late last March when I discovered fresh digging there. After that, I went daily and stealthily to the Far Field.

Then early last April, as I stood gazing across the old field at the den, an adult red fox emerged from the hole. Luckily the wind was in my favor and the fox never saw me. It sat down and scratched itself, then stood up and looked around. Finally, it poked into the den exit and I caught a movement. Could it be kits? I was not to know that day because as suddenly as the fox had emerged, it slipped back down into the den.

The following day, as I rounded the first curve in the Far Field Trail, I came face to face with a red fox, probably out hunting food for its family. It veered abruptly and bounded off in a zigzag pattern, while I continued on to the Far Field. Walking slowly down the edge of the field, peering through my binoculars at the den site, I spotted an adult fox sitting outside the den, only this time it was accompanied by two handsome kits, still wearing their charcoal gray, natal coats. Quietly I sat down in a black locust grove beside the field edge to watch the little family and soon verified the sex of the adult when the kits began to nurse. As I steadied my binoculars on my knees to better observe the action across the hundred-yard field, the kits pulled away from the vixen and started exploring by climbing over fallen branches in the vicinity of their den. The vixen, who alternately paced and reclined, watched them, and once she stopped to groom a kit. I remained glued to the spot, unwilling to move even when it began sleeting, but at last they all went underground and I walked home in the worsening weather, elated to have finally discovered an active fox den with young kits.

I did not see them every day, but with each sighting I learned a little more about the family. At ten o'clock one morning, I found the female outside with *four* kits. They played with each other and with her, and although she moved around occasionally, she seemed to be unaware of my watching, hidden as I was by the locust grove. Later, in the early afternoon, the kits, guarded by the vixen, poked about in the brushwood outside the den, jumped on each other and sometimes on their parent. Then the dog fox trotted up and lay down and the vixen retired to the den, while the kits continued playing for another fifteen minutes until they also went underground.

By the sixteenth of April, the kits were still charcoal gray, but already they were learning about food. One of the adults had a carcass in its mouth which it offered to any kit that came near, but although each kit would make a feinting movement as if to grab it, it would always back off at the last minute. Apparently, it had been designated as a plaything because they never did eat the carcass and, after a while, they ignored it altogether, seeming to prefer exploring further uphill or tussling with each other. Scientists will say that they were working, learning how to live in a complicated world, but to me it seemed like children's play—a good deal of fun with some learning mixed in.

To my distress, that was my final look at the foxes last spring. For weeks I feared that they had been killed, since red foxes are hated by many people. Although foxes occasionally eat the creatures—rabbits, ruffed grouse, wild turkey—that humans want to hunt, smaller prey—chipmunks, meadow voles, and insects, as well as wild fruits and the carrion of larger animals such as deer—still forms the core of their diet. After several days of searching, however, I found a second occupied fox den in a secluded grape tangle less than a quarter mile from the Far Field den, so I was hopeful that the little family still survived. Foxes often have more than one den and move their kits back and forth depending on conditions.

So this year, knowing that foxes mate for life and that vixens usually reoccupy the same den, I am looking for the telltale double tracks of a male and female traveling together which I know signal the onset of red fox courtship. I also visit the Far Field den for signs of occupancy. To my delight I discover an adult red fox scratching itself as it lies just outside the old den entrance hole. Then it settles back to snooze in the sun. I ease myself over to the melted edge of the field, slip off my snowshoes, and sit down on the snow, cushioned a bit against the cold by the hunter's "hot seat" I carry hooked on my belt during the winter months.

I peer through my binoculars at the recumbent fox, but all I can see is a motionless, red, furry ball since the fox has its head tucked under its chest. Even when a flock of crows fly noisily over, it does not stir.

Eventually the cold drives me to my feet, and slowly I strap the snowshoes back on and move down along the edge of the field. The fox sits up alertly and stares in my direction just as two ravens croak their way over the ridge. Silently the fox and I eyeball one another across the field. Then the fox rises to its feet, lopes up the hill, veers sharply right, and disappears over the far crest of the field.

What a wonderful, long view I have of that beautiful animal running over the snow. Its burnished red coat against the white and black of the woods is the most colorful sight I have seen since the autumn leaves fell. Visions of fox-watching dance in my head. This spring will be even more wonderful than last, I tell myself.

Three days later a warm wind comes in overnight and by eleven in the morning it is sixty degrees on the back porch. Crows caw steadily across First Field while a pair of ravens sport with the wind. The singing titmice, cardinals, and song sparrows are joined by the pensive "poor Sam Peabody, Peabody, Peabody" of the wintering white-throated sparrows.

Later a tumultuous downpour makes a start in clearing the mountain of snow, and when I step outside I am suddenly enveloped in blackbird noise. One hundred red-winged blackbirds have landed on two locust trees outside the house. I stand transfixed, absorbing the first, tentative "okalees" of the returning males. For many years we had one breeding pair that made a nest in First Field and brought its youngsters to drink at the driveway ditch, but normally they will not remain to breed because our mountain does not provide proper swampy habitat for them. Since I never banded the ones that stayed, I can only wonder if they were a family that retained, year after year, the ancestral memory of our place as home and if they had finally all perished, leaving no red-winged blackbird with any knowledge of our mountaintop.

Every February, when the red-winged blackbirds visit, we hope that another pair will establish a family here. The boys even planted cattails in the low, wet area of First Field as an inducement several years ago. The cattails have spread, but the blackbirds have not taken the bait. Studies show that red-winged blackbirds are the most abundant bird species in all of North America yet they, like the other so-called pest species— the house sparrows, common grackles, and starlings—refuse to breed here. I, for one, miss their calls, so reminiscent of wet, mysterious places, and eagerly await the spring when once again I can hear them on a daily basis.

By late February the striped skunks are courting. Although they retreat to communal dens in what the scientists call "carnivorean lethargy" during the most severe weather, the males are out and about as soon as the weather moderates.

The males are polygamous and travel widely in their quest for females. They are also liable to be a bit out of sorts. Competing males growl and claw at one another and sometimes even release some of their noxious spray in the excitement. Usually I catch my first whiff of "woods pussy" perfume at the top of First Field. For the next couple of weeks the males will wander from den to den in search of receptive females, all of whom come into heat for a period of four to five days and are as polyandrous as the males are polygamous.

Mating is a rough affair. The male unceremoniously grasps the female's neck with his teeth and mounts her from the rear. Once the matings are over, the female builds a den which usually consists of two, twelve-foot tunnels leading to a chamber three feet underground. She lines it with up to a bushel of dried grass and, after a sixty-three-day gestation period, gives birth to four to six sparsely haired, blind and scentless kits. It is two weeks before they grow hair and still another week until they open their eyes. At that time they are able to assume a defensive posture and emit a small amount of noxious scent.

Their mother takes scrupulous care of them, keeping them clean and weaning them at between six and eight weeks of age. Then the kits are ready to follow her around on evening hunting forays. While some finally leave their mother's care at the end of the summer, others may stay with her until the following breeding season.

Those who have kept pet skunks have found them to be endearing creatures. Yet most people fear and hate them because of their overpowering defensive scent. Even their scientific name, *Mephitis mephitis,* means "noxious odor." Since skunks are found only in the Western Hemisphere, early explorers were amazed at such creatures and decided that the smell they emitted had something to do with their excretory habits. The highly sulfuric n-bulymercaptan chemical is contained in the anal glands, so this mistaken idea is easy to understand. On moist days, the odor can be smelled more than a mile away, but surprisingly the oil in the glands has been refined and used as a fixative in perfumes.

Of course, the skunk does not use its ultimate defense weapon unless it is sorely pressed. First, it stamps its front feet as a warning, then it rudely turns its back, lifts its tail, and finally shoots—accurately aiming for the victim's eyes at a distance of fifteen feet. Temporary blindness, for up to twenty minutes, can result, but washing the eyes with water will take care of the problem.

Whenever I have run into a skunk, I have been struck by its total lack of concern with me. Actually, skunks have such poor eyesight and such confidence in their defense mechanism that they walk along, nose to the ground, more interested in finding food than watching out for predators. During one mild morning in February, I noticed a skunk moving quickly along the base of Sapsucker Ridge and I paralleled its path at a distance of less than twenty feet until it disappeared into heavy brush, its preferred habitat. I don't believe it knew I was there. Last summer one walked directly across my path less than ten feet in front of me as I stood frozen in place.

Few animals prey on the skunk. Only great horned owls

and barred owls, with poorly developed senses of smell, find them delectable. Otherwise, humans are their principal predators, poisoning, trapping, shooting, or gassing them in an attempt to keep them away from their homes and beehives. Skunks appear to be immune to bee stings, and in the years when we kept a few hives we did see evidence that a skunk had visited occasionally. But their favorite insects are mostly those which we would like to get rid of—gypsy moth caterpillars, potato bugs, Japanese beetles, crickets, and grasshoppers. Other popular foods in season are wild berries, small rodents and rabbits, dried fruit drupes, carrion, and garbage.

Altogether the striped skunk is a creature to be enjoyed for its beauty, which is what I do whenever I am lucky enough to see one.

In late February the male woodchucks are also out in search of females, and every woodchuck hole in the fields and along the Far Field Trail has fresh tracks going in and out.

This is also the time for our annual visitation by American kestrels. I say visitation rather than return because after several years of raising families in the power pole halfway across First Field, the American kestrels have only been paying us brief, early spring visits before settling on valley farms to nest. But during their February visit, the male's "killy-killy" rings from the field, and there, on the top of the old nesting pole, he sits and surveys the scene around him. When I walk up the field toward him, he flies to the edge of the woods where he lands on a tree branch and calls several more times. As I approach still closer, he flies across the field, calling loudly, and alights on the topmost branch of a white pine tree.

At precisely the same moment, a female comes swooping down over the power line right-of-way from the Sinking Valley side, answers the call of the male, and flies over to the power pole where she lands for a few seconds. Then she moves over to a second power pole which she touches down on briefly. Finally she flies to the male in the white pine, hovering over him as if she is greeting him.

At last she spirals high up in the sky where I lose sight of her. After a few minutes the male, still calling, flies back across to Sapsucker Ridge. Although most of these pretty little falcons, once misnamed "sparrow hawks," migrate to the southern United States and northern Mexico in the fall, some do winter in pairs in the northern states. I have no way of knowing whether the pair I saw are early arrivals or have been wintering in the valley, but hearing their calls does bring back memories of their nesting days here when the whole family coursed noisily back and forth over First Field on the hottest days in late July and early August.

Finding chipmunks out and about is still another late February discovery. They always choose a mild day to replenish their stores and practice their "cuck-cuck" calls. Chipmunks are not true hibernators because they waken every several days to eat from their store of seeds and nuts cached in their underground dens. The only month I have not seen chipmunks at least once in the woods is January; a few of them nearly always appear whenever the weather moderates. So on this last day of February, chipmunks appear along with warm sunshine and a blue sky studded with clouds racing along in a pre-March breeze. I can step now from winter into spring by walking from Laurel Ridge, still covered with four inches of frozen snow, to the First Field, Sapsucker Ridge, the Far Field Trail, and the Far Field itself, all of which have warm southern exposures where much of the snow has already melted.

I am drawn to the base of Sapsucker Ridge by what sounds like a mixed flock of blackbirds and pine siskins. But all I find are eighty or more pine siskins running over the snow-free ground, giving their usual goldfinchlike calls as well as the blackbirdlike, buzzy sounds I first heard. Occasionally they swoop up into saplings in response to warnings I cannot hear or see, and I sit to watch as they resume their running and pecking in the ground. White-breasted nuthatches and downy and hairy woodpeckers are also climbing up and down the saplings. I am surrounded by birds going about their business and ignoring my silent, bemused presence; some come within ten feet of where I stand watching. After a half hour I move on, leaving them to their exuberant work.

Later I sit at the edge of the Far Field when a fighter plane suddenly screams low across the field, barely skimming the treetops. In fact, from where I sit, it seems to be flying through the trees. Its flight disturbs a red-tailed hawk perched on the ridge and it flies up out of its cover, heading in the same direction as the jet, looking like it is being pulled into the plane's wake.

Red-tailed hawks are also mostly migrants, and that is the second sighting I have had of one this week. To add to my bird-of-prey sightings I watch both a Cooper's hawk and two turkey vultures flying along Sapsucker Ridge this morning. Our ridge is one of the migrating bird-of-prey ridges both in the fall and spring since it runs for hundreds of miles north to south, the northernmost ridge in the ridge-and-valley province of the eastern United States.

Such a day of warmth and chipmunks, pine siskins and birds of prey has set the stage for the event which is about to begin—the debut of a true Appalachian spring.

MARCH

Overture

MARCH 1. March nearly always arrives like a lamb here on the mountain, and today was no exception. It beguiled me with promises of what is to come, and I thought, prematurely, that spring was here to stay. So of course I performed one of my first spring rituals. I put up my spring arrivals' list on the refrigerator door.

I started such a list when we lived in Maine many years ago, inspired by the arrival of a singing eastern meadowlark. My list then grew as more and more birds returned and wildflowers blossomed. I recorded the arrival of the burbling bobolinks, the day the ice finally went out on the lake, and what date the white trilliums bloomed in the pine forest.

Alas! We have no breeding bobolinks or meadowlarks on the mountain, but we do have indigo buntings and field sparrows. Instead of white trilliums, purple trilliums bloom along the stream in large, drooping patches. Rufous-sided towhees, which had not yet reached central Maine when we lived there, are among the most common birds on the mountaintop.

It took me several years to redefine my list once we moved to the middle latitudes. Now I have a list of birds and flowers that stretches back to 1971—forty-one items including such notables as wild azalea, bluebird, wood thrush, trailing arbutus, and northern oriole. It has been redrawn and figured ahead by our poet son, David, and will not run out until 2001. It gave him, he told me, a wonderful sense of faith in the future when he did it.

For years, when the boys were all home, they vied with each other to report sightings. My list was the favorite topic of conversation for three months of the year, and if I forgot to put it up when they spotted the first red-winged blackbird, I was properly chided. Even now, whenever they are home in spring, they check the list right after they say hello. Nature friends from afar who come to visit have been known to stare amazed at what I choose to adorn my refrigerator with.

When David revised the list, he alphabetized it, beginning with "Azalea, wild." A quick glance shows me that April 28 (in 1977) was its earliest blooming date and May 23 (1984) its latest. Probably because we are so eager for spring, a disproportionate part of the list is devoted to March happenings. And even February appears a few times. Back on February 26, 1978, for instance, an amorous woodchuck surfaced on top of the snow just below our back porch, and we watched him plodding along for several minutes after which a heated argument ensued. Was he or wasn't he a spring arrival? I buckled under the intense pressure to so record him, knowing full well that his emergence was temporary and it would be at least another five weeks before any self-respecting woodchuck would be permanently awake. In fact, the previous year we had not seen a woodchuck until May 7.

In 1981 a garter snake stretched itself in unseasonable warmth on February 21; Canada geese have flown over as early as February 19. Robins and song sparrows, when a few have not wintered over, have appeared as early as February 26 and February 19 respectively. Eastern bluebirds too are sometimes February arrivals, and in 1984 they first appeared on our telephone line by February 22.

But by and large the bluebirds come in March, along with northern flickers, eastern phoebes, mourning doves, and field sparrows. The only wildflowers that may bloom before the end of the month are coltsfoot and trailing arbutus.

As avid birders will testify, listing can be an exciting pastime. Searching for new arrivals and blossoming plants sends me out in all kinds of weather and gives me a more intimate relationship with the new season. Because of the vagaries of early spring weather, the list does not help me to predict whether the season will be early or late. It merely buoys my spirits during setbacks and records, in a rough way, the temperament of a particular spring—premature or tardy, benign or severe.

Yet whether or not the phoebes have arrived as early as March 10 or as late as March 26 or the wood frogs have begun courting on March 16 or April 5 probably doesn't matter in the long run. What does matter is that the phoebes have returned as usual and the wood frogs have courted as expected. In a world full of uncertainties, it is comforting to know that I can still depend on the certainty of spring.

MARCH 2. Sunbathing days are here again, and I dragged the old chaise longue out to the veranda this morning once the thermometer rose to sixty degrees. I heard them then, the soft warbles that indicated eastern bluebirds. After a quick scan I located a male and female in their usual place on the telephone wire just above the wooden bluebird box I erected four years ago.

Less than a decade ago bluebirds were occasional birds of passage both in spring and fall, but now they nest here. In fact, last year I heard their soft calls during every month but January. There is no doubt that bluebird numbers are again increasing; for once humanity has set out to save rather than to exterminate a species. By mounting thousands of nesting boxes in suitable habitat all over the eastern United States and

Canada and trying to protect the birds from more aggressive bird species, such as starlings and house sparrows, as well as from predators and parasites, humans have helped to set the bluebirds' numbers on the upswing. As a result, what used to be uncommon is becoming common once again, at least here on our mountaintop. But no matter how often I see them I never cease to marvel at the incomparable bluebird blue color of the male's back and head.

I watched as they fluttered down into the field several times after food. Finally the female, swooping low over the field, flew off. The male landed in the old Seckel pear tree beside our driveway and spent several minutes preening his feathers. From there he proceeded to the large staghorn sumac bush where he was joined by a second male. Both ate the dried, fuzzy, red berries that are favorite winter and spring food of bluebirds, no doubt unaware of the brilliant contrast between the blue of their backs and the red of the sumac heads. It remained, for me, an unforgettable image of spring beauty.

MARCH 3. Brown-headed cowbirds have had a bad press because of their propensity to lay their eggs in the nests of other bird species who not only do not recognize the eggs or resulting hatchlings as strangers, but will feed the pushy youngsters to the detriment of their own blood relations. The sound of their liquid song in a still-silent woods is, nevertheless, a welcome sign of spring. And to watch several males lined up beside a female, giving what bird behaviorists Donald and Lillian Stokes call their "topple-over" display in an effort to win the female for their own, is one of my earliest spring entertainments. Today, in the black walnut tree in my back yard, they performed the ceremony for the first time this spring.

As they sang, several males fluffed up their feathers, arched their necks, spread their tails and wings in unison, and then fell forward, catching themselves up just as they appeared to be headed for the ground. This seemingly peculiar behavior apparently stimulated the female, and in some indefinable way she eventually chose what she considered to be the best performer to sire her offspring. Once the choice was made, the victor spent his time continually pushing other would-be suitors away from his chosen one, like the biggest kid on the block elbowing out all contenders.

MARCH 4. Spring has retreated—as I knew it would—and the successive rain, sleet, and inch of wet snow that fell last night put a sheen on the mountain. The slippery elm sapling below our back porch sparkled in the occasional bursts of sunshine, and dozens of small birds paused on its bright branches before coming into the bird feeder. I went out to listen to the mountain ringing with birdsong.

Black locust and Carolina poplar branches littered the yard. The ice gave the poplar a golden glow, but the black limbs of the locusts, cast in heavy silver, looked like Aztec treasure. Hanging from each tree limb were frozen droplets, some longer than three inches, that caught beams of sunlight against an Alpine blue sky. At least an inch of snow lay atop the limbs. As I looked directly into the sun through the icy trees, my eyes were dazzled by the crystalline light. But when I gazed down the line of icy trees rimming First Field on which the sun was obliquely shining, I caught the twinkle of miniature Christmas lights—blue, red, green, yellow, and orange—scattered amidst translucent droplets sparkling in the tree branches.

From the edge of the field I watched two bright-red male cardinals chasing each other in the silvered woods while the female looked on. A drumming hairy woodpecker set up a percussive counterpoint to the delicate sound of tinkling ice crystals that shattered whenever the slightest breeze whispered over the ridge.

Every tree, every shrub, every weed was a unique sculpture shaped by ice, and I was overwhelmed by the multitude of artistic creations. My only regret was that my footsteps sounded like thunder as I crunched through the ice. To stop their reverberating echoes, I sat down in the midst of broken ice shards and ice-pruned branches. Silence enveloped me and I was struck by my solitary enjoyment of the scene. How many similar scenes of natural splendor play to no human audience?

At last, thoroughly satiated by beauty, I turned homeward, my way still lit by prisms of light that flashed the colors of the rainbow.

MARCH 5. They call it *chinook* in interior British Columbia, that warm, damp wind from the Japanese current off the Alaska Panhandle that gains further warmth by losing part of its wetness in the Coast Range. When it sweeps down the eastern sides of those mountains it changes thirty-degree-below-zero temperatures to fifty-degrees-above in a matter of minutes.

Snoweater was the local Indian name, which is exactly what happened on our mountaintop today. At 6:30 in the morning a warm breeze blew from the south and our thermometer stood at sixty degrees. In the valley it was still forty degrees, but within a short time the warm wind had blown down our slopes and raised the temperature to a balmy seventy-five. And with that dramatic event we returned to spring in a few hours.

I had to be away today and left a white and silver mountain in the morning. I returned to a brown one this evening and a report from Mark that the bluebirds had appeared again.

MARCH 6. A day of warmth, of beauty, of geese and swans. At 9:30 A.M. I heard the first contingent of geese fly over just as the sky, within fifteen minutes, swept itself clear of heavy clouds and became infinite blue. Our birder son, Steve, went racing out with his spotting scope, settled himself on a rise in the middle of First Field, and began counting as flock after flock of Canada geese and tundra swans streamed over the mountain in wavering V-shaped formations that ranged from 40 to 300 birds.

The finale of the show occurred at 11:30 A.M. when it sounded as if the sky would collapse under the burden of noise from calling geese. As we gazed up, squinting against the strong sunlight, we saw to our left a flock of 40 geese, to our right a flock of 120, and in the middle, at the rear, a flock of 300. Our eyes, our ears, and our emotions were totally caught up in the pageant of migrating waterfowl. Steve's final tally for just two hours was 2,420 Canada geese and 440 tundra swans. Never, in all our years here, had so many geese and swans passed over in such a short time.

MARCH 7. This morning it was forty degrees and overcast with just a hint of snowy sleet coming down, but in the dank silence I heard what felt like a tom-tom beating in my head, a sound that baffled me during our first years here—a male ruffed grouse drumming. Along the Far Field Trail I found both the drummer and his log, but he slipped off without giving me a show. Although ruffed grouse will occasionally drum at other times of the year, since the drumming also delineates a male's permanent territory, it is primarily a device for attracting females. For that reason the spring drumming is more determined and persistent, a definite sign that spring is here despite the wintry weather. As if to verify that, a flock of tundra swans swept over the ridge heading north just as the grouse exited.

MARCH 8. Steam rose from the frosted earth at dawn, but as I stepped outside to say good-bye to Bruce at 7:30, my ears quickly tuned in the sounds of calling geese, a singing bluebird, and the "tut-tuts" of robins. Up on First Field I spotted an enormous flock of robins running, probing into the thawing earth, and flying low back and forth over the field. An hour later they were still there, along with five starlings and one common grackle. Both the starlings and robins ate sumac berries in addition to whatever nourishment they were obtaining from the ground.

I sat in the field surrounded by robins and warmed by the morning sun. A single junco trilled from the top of a black locust tree at the edge of the field. Hairy woodpeckers courted, bluebirds sang without ceasing, crows cawed, and cardinals "cheer, cheered, pretty, prettied." New, vibrant life seemed concentrated on the upper, warm fringes of First Field.

Then a new sound entered my consciousness. For the first time since last summer I heard the keening cries of cedar waxwings. First a single one, then two more flew in, their golden breasts shining in the sun, and landed on a nearby black locust sapling. I thought that they too had intended to eat the sumac. But before they had a chance, a red-bellied woodpecker displaced the robins and starlings and sent the waxwings whirling away across the field and out of sight with its maniacal cry.

I walked on across the field toward Sapsucker Ridge, accompanied by the cry of geese overhead, but the day was so clear that they were flying too high to be seen. Then, as I moved along the Sapsucker Ridge Trail, I both heard and saw thirty tundra swans, shining white against a blue sky.

At the Far Field I was greeted by the song of a bluebird and the calls of pine siskins, and as I stood listening, a mixed flock of blackbirds streamed past, although that field had not attracted the large numbers of birds which First Field had. But then spring always begins on First Field, which skitters with tiny spiders as soon as it starts to warm up. Those spiders provide good protein easily caught by migrating birds.

MARCH 9. I rarely find what I expect to when I go walking on the mountain. Instead, it is the unexpected I almost always encounter. Take, for instance, this damp, overcast morning. I was walking along, listening hopefully for spring birdcalls, when I noticed a large, dark, round shape in the crotch of a tree—a shape I was quite certain I had never seen there before. Through the binoculars I could tell it was the rear end of a wild creature.

At first I thought it might be a small bear. But as I walked closer, I could see its tail hanging down, sporting stiff black bristles tipped in white. It was definitely a good-sized porcupine.

I scrambled down the slope to get a front view of the creature. It knew I was there because it moved its head with sloth-like slowness as I came closer. Still, it made no attempt to escape, even when I stood beneath the tree and looked up at it. Instead it stared myopically down at me. I had probably disturbed its rest because it had done no feeding at all on that particular tree. I did notice that one branch on a neighboring tree had already been debarked. Obviously the porcupine was still enjoying his winter diet of tree cambium and phloem.

This is not the first porcupine I have seen on the mountain, although my sightings over the years have been few enough to excite me when they do occur. Usually porcupines are most visible in the winter when they tend to spend days in one tree, eating its bark.

As I stood looking at the impassive, seemingly dull creature, I couldn't help recalling our Plummer's Hollow porcupine. Of all the porcupines I have seen, it was the only one to display an interesting personality. It had been wintering in our hollow and had even allowed Steve to climb up next to it in a hemlock tree and draw it while it gnawed at the bark.

Then one cold evening I had hiked down the road to meet Bruce. The porcupine had just descended from its favorite hemlock tree, so I walked right up to it to get a close look. It scrunched up into a small crevasse in the tree trunk, tucked its head under its body, and erected its quills; but it made no threatening noise or motion. After all the antiporcupine propaganda I had read, I was surprised by its small tail and its unaggressive manner. I spoke a few quiet words to it before moving on.

After that I often walked down to the porcupine tree to watch it, although it was always high up, sitting very still or showering the branch tips down as it ate. The boys, too, reported on its actions each day when they walked home from school and even brought friends up to show them the porcupine in the hemlock.

The porcupine has been universally despised by humans because of its peculiar manner of defense. John Joselyn in 1672 first described it in his ground-breaking study, *New England Rarities Discovered,* as a "very angry creature and dangerous, shooting a whole shower of quills with a rowse [shake] at their enemies which are of that nature that whenever they stick in the flesh they will work through in a short time if not prevented by pulling them out."

The quills do work through the body by the muscular contractions of the victim, reaching vital parts and killing the creature if they are not pulled out, but porcupines do not throw their quills when they are attacked. The quills are loosely attached to their bodies and come out easily when they swish their tails in defense. The word *porcupine,* in fact, means "quill pig" in Latin and it is also called "porky hog," "prickle pig," "quiller," and "silver cat."

Because they have a craving for salt and eat the inner bark of trees—which sometimes kills the tree, porcupines are generally shot on sight around here and anywhere else that they live. Yet those who have come to know them best are lavish in their praise. One researcher, Dr. A. R. Shadle (according to Doutt, Heppenstall, and Guilday in *Mammals of Pennsylvania*), reported on what he called their "exercise dance," when they stamp about on their hind legs with a definite rhythm, and Ronald Rood, in his charming book *How Do You Spank*

A Porcupine?, described his pet porcupine's dancing. He also stressed the intelligence and charm of the animals and emphasized that porcupines only seem stupid because their eyesight is so poor. He maintained that they learn quickly, depending on their strong sense of smell to locate food and on their acute hearing to warn them of danger.

One evening Bruce had to stop the car, get out the snow shovel, and gently scoop the porcupine off the road in the hollow. The next morning it was again blocking the car. This time Bruce took a stick to carefully prod it; the porcupine grasped the stick in its front paws and pulled away. It made no threatening move and showed no alarm. It seemed, in fact, to be getting used to the strange creatures it kept encountering in its territory. Bruce had to edge the car cautiously around the resolute animal.

That afternoon Steve came home cradling the porcupine in his arms. "Someone shot him," he sorrowfully reported. "I found his body in the road." We were appalled and saddened by the senseless death of our charming and peaceable friend. Despite its quills, it was defenseless against a person with a gun. When I later talked to a friend about what we considered a cowardly attack on an inoffensive wild creature, she informed me that many people will go out of their way to kill porcupines because they consider them useless creatures that are a nuisance to humans and their dogs.

I wondered then what kind of a world we would have if people killed everything they believed useless to them. Must all wild creatures be useful to humans to justify their existence? Sadly, in many cases, it is so. Witness, for instance, how much money has been spent on managing fish, birds, and animals for game hunters and in hiring professional hunters or paying bounties to rid the world of all the creatures, such as coyotes, that compete with humanity.

Since the shooting of that porcupine, I have never again had more than a passive look at one. So it was with today's porcupine. I sat with my back against a nearby tree for over half an hour watching it, and during that time it did nothing but return the favor. With such an impasse of inaction I once again had to be content. Finally, I left it to its own slow devices. While watching a porcupine may not be high-level excitement, it was enough to make my day, which goes to show that Clint Eastwood and I definitely exist on different planes.

MARCH 10. Steve always tells me that I must look up as I walk, but this morning my head was down and I was lost in thought. The cry of a gull over First Field abruptly jerked my head upward. Fifteen ring-billed gulls mewed and circled over Laurel Ridge while twenty-five more split into two groups, one of which circled clockwise, the other counterclockwise, over Sapsucker Ridge. I closed my eyes, sniffed the air, and imagined that I could smell the salt air and hear the waves breaking against the shore even though I know that ring-billed gulls are as much at home on inland lakes as they are at the seashore.

Every spring the local papers make much of seeing a flock of gulls fly over the town, intimating that they are far off course. In fact, they are heading for the lakes of the northern United States and southern Canada where they breed on islands and eat very ungull-like food such as small rodents, grasshoppers, worms, and grubs along with the more usual fish and garbage that their larger relations, the herring gulls, scavenge. Although they closely resemble herring gulls, they can be easily distinguished from them by the black ring near their bill tips and their yellowish or greenish legs. Whenever we go to Washington, D.C., in the winter we always stop on the mall to watch the hundreds of ring-billed gulls that winter there, striding about and giving the place a seaside air. When we hear them crying overhead on our mountain in March we count them as spring flyovers, another sign that our northern spring is progressing as usual.

MARCH 11. Despite snow flurries today, another avian visitor stopped in for a scratch beneath the bird feeder. The handsome fox sparrow is larger and more gaudily dressed than our usual sparrow species, all of which Bruce consigns to "dicky bird" or LBJ (Little Brown Job) status. But even Bruce, who is incredibly obtuse about seeing the subtle differences among song, field, tree, and chipping sparrows, recognizes that the fox sparrow stands out from the "dicky" crowd with its foxy red back and head, its superior manner, and its vigorous scratching by kicking backward with both feet.

Although I have frequently come across small flocks of fox sparrows at the Far Field in the fall, they always appear singly here in the spring and always in the grape tangle where there is spilled bird seed and protection from predators and the elements. This one stayed only for the day—eager, I imagine, to head on up to Canada where many of them breed.

MARCH 12. Spring nights too have their rewards, especially when it is clear. Bruce and I had the rare privilege of going out and seeing both the winter and summer constellations sharing the sky at 9:00 P.M. Bruce, as he has been doing for nearly twenty-eight years, pointed out all the constellations, and I, as I have been doing ever since he first started the fruitless exercise of trying to teach them to me one night while we were canoeing on Eagles Mere Lake, nodded enthusiastically and tried to convince him that I really did see them all.

Before I met him I could recognize Orion and the Big Dipper. Since then I can definitely pick out Cassiopeia and sometimes I may recall the Seven Sisters, but I generally find the constellations impossible to tell apart. To me, stars are only themselves—awe-inspiring, beautiful, but not of this world. Whenever I think of how vast their distance is from earth, I am humbled. And I wonder how humanity can be so self-important in the face of such infinite unknowingness. Even the most brilliant scientists seem to learn less and less about more and more the deeper they delve into the universe.

The lesson, by now, should be learned. As far as we know, we are alone in a universe where only our small, finite earth can support us. To save the earth from destruction should be our primary concern. Nothing else matters if we destroy earth's ability to sustain life. Yet to listen to the prate of politicians this election year is to despair. Pettiness and mudslinging is the fashion, and politicians everywhere seem to fiddle while the earth burns.

MARCH 13. Song sparrows are almost always with us, but March brings them in to pack their breeding territory as tightly and as early as possible. Despite the weather, which today was hazy and cold, they all proclaimed, "Hip-hip-hurrah, boys! Spring is here!"

While each song sparrow is an individualist with his own version, the song is usually recognizable as a song sparrow song. But today I heard the most ghastly croaking and thought that we had a new bird species in residence. Never had I heard anything quite like it. I spent a good deal of time stalking the tuneless singer and finally discovered that it came from a male song sparrow establishing territory in the grape tangle. I was amazed! Had I not seen it with my own eyes and heard it with my own ears, I would never have believed that such a noise could emanate from the pipes of a song sparrow. Could he have been born without proper vocal cords?

MARCH 14. In the warmth of the morning sun, I caught the flutter of a butterfly in the barnyard and went down to watch as it landed on the warm, white side of the barn and spread its wings. It had a fat, reddish brown body whose color radiated out to the middle section of both wings. The rest of the wing area was brilliantly marked with splotches of black, orange brown, and white. A thin strip of yellow beige along the wing edges finished off this beautiful creature. It was a Compton tortoiseshell, in the same genus *(Nymphalis)* as that other early butterfly—the mourning cloak.

Both species winter in adult form in hollow trees, old barrels, boxes, and tin cans and will emerge during warm spells. As I watched, it flew swiftly about in the sunshine, but it frequently paused to bask in the heat radiating from the white barn siding. Once when it alighted, it folded its wings, resembling a dead leaf with its ragged, brown, upper wing edge, a wide band of gray in its middle, and more brown on its lower wing edge.

The Compton tortoiseshell is a northern, woodland species which is attracted to tree sap, rotting fruit, and mud puddles. Its favorite food trees are northern white birch and willows, neither of which we have around our farm. We do, however, have poplars—which the field guide says they "perhaps" like. The Compton tortoiseshell is noted for its periodicity, which means that some years it is abundant and other years it is nearly absent. Perhaps this will be a good year for them.

MARCH 15. Every time I hear the nasal "peent" of a woodcock, I regret anew that we do not have ideal woodcock nesting habitat. Except for one spring twelve years ago, we have never witnessed the courtship flight of a male woodcock. Yet I have never forgotten that magic evening when Steve came running inside at dusk shouting, "The woodcocks are courting." Hastily we went dashing out behind him and up the driveway to stand and listen.

First came the nasal "peent," then the tumbling song from the sky, and finally the whistle of his wings as he plummeted toward the ground. Over and over the cycle was repeated, but we never found either the hidden female or the male himself on the ground because it was too dark. Nevertheless, the music and grandeur of the night was enough to satisfy us. Little did we know as we listened that it would be our only chance to hear the courtship flight of the woodcock on our mountain. Since then it has remained merely another bird of passage which rises abruptly in front of us in some wet area on a mid-March day, its disproportionately long bill its unmistakable badge of identity.

MARCH 16. The first eastern phoebe arrived silently, as he always does, and spent the day on the electric wire beside the barn roof, hawking insects off the sunny side of the barn. It will be several days before he ventures a song, but with the return of the phoebes, spring is well on its way. Of all the migrating birds that make their homes here it is the phoebe who comes the earliest and stays the latest, until the third week in October. Often we have four families in residence at the same time, and each feels honor-bound to have two broods. So we have the guesthouse portico phoebes, the springhouse phoebes, the outhouse phoebes, the garage phoebes, and sometimes, for a second brood, the veranda phoebes.

But I can never have enough of phoebes. They may not have a beautiful song, in fact it is downright monotonous, but they are exemplary parents and they move in so close to us to raise their families that we cannot help becoming fond of them and of their numerous nestlings.

MARCH 17. A cold wind swept across First Field this St. Patrick's Day, but Bruce and I set out on a late morning walk and were instantly cheered by calling robins hunting for food near the top of the field. As we headed toward the birds, across closely shorn grass still dotted with patches of snow, we suddenly realized that the robins were not alone. Much smaller, brownish birds ran along the ground, partially obscured by tiny hillocks, and even my binoculars could not bring them in close enough for me to identify. I was almost certain, however, that I was seeing a new-to-the-mountain bird.

Excitedly I asked Bruce to return home and retrieve our Peterson field guide while I kept an eye on the birds. I also searched my memory for a clue to the birds' identification. Occasionally I can pull the name of a new wildflower, bird, fern, or tree out of the air before I check any book. The name I summon up out of my subconscious almost always turns out to be correct.

So it was with the meadow birds. "American pipit" sprang to my mind even before I stalked them and noted their undistinguished brown backs and almost white breasts streaked with light brown. When Bruce returned with the field guide, it confirmed my guess. As I watched them, they continually wagged their tails up and down whenever they paused. If I came too close they would fly up in a cloud, flashing their pale breasts as they bounced up and down, all the while emitting high, piping calls.

American pipits, *Anthus spinoletta,* are circumpolar birds of the cold, windy tundra. Their winters are spent in warmer climes—southern Asia, central America, and northern Africa. In the United States they sometimes winter as far north as Washington state, northern Ohio, and Massachusetts. But most of them prefer our southern states where they can eat a wide variety of insects and weed seeds.

They are one of the earliest species to start north in the spring, headed for the tundra of arctic Canada, the barren coast of Labrador, or the treeless slopes of our western mountains. There they court and build their nests of grass and twigs on moss-covered hummocks. The female broods the four to seven pale eggs heavily marked with dark brown while the male feeds her and later helps with the care and feeding of their young.

Sometimes American pipits winter in our area and can be found in the valley fields where the grass is short or the land has been recently plowed. They also like to search for small invertebrates along the unfrozen edges of lakes or ponds. But the fifteen pipits that spent the day on our meadow were probably migrants, enticed by the sight of our closely cut, hillside meadow so reminiscent, with its snowy patches, of the high tundra world they inhabit. Here, as there, they could run on their long dark legs and poke into the thawing earth for insect food.

I followed them around for several hours, observing their behavior through my binoculars. Often I sat still on the cold ground for long periods. Gradually, then, they would come close enough for me to catch the subtleties of their coloring. I could clearly distinguish the light streak above each eye, the brownish gray of their upper backs and the darker gray and brown shading on their wings. Sometimes they scratched with their feet or stopped to preen their tail feathers, spreading them wide enough so that I could see the white lines bordering the dark gray. Watching them most of that long afternoon, as they ran over our barren ground still patched with snow and swept by wind and cold, I felt as if I had been magically transported to the tundra for a few hours.

Near dusk the pipits rose into the air for the last time and piped their farewell calls into the brooding silence. Will they ever come again to our mountain, or was their visitation on St. Patrick's Day, like the woodcocks' courtship, another once-in-a-lifetime experience?

MARCH 18. Many of spring's happenings occur underground, far from the eyes of watching humans. It is there that most of our mammals court and breed and bear their young. No wonder so many more humans are bird- rather than mammal-watchers. Most birds are far less secretive in their amatory pursuits.

By now male woodchucks have thoroughly awakened from their hibernation and are on the prowl for females who wait in their dens for males to come calling. Two to five young will be born between mid-April and mid-May—the gestation period for woodchucks is a mere thirty-two days. Although they are born naked and helpless, they are weaned by six weeks and on their own at two months of age, just when gardens are ripe for raiding.

Woodchuck dens are used by several other species as well. Sometimes several female and immature striped skunks share a den with a female woodchuck! Or they take over deserted dens. Red foxes, for example, rarely build their own dens, but let the woodchucks do the work. By now the fifty-one-day gestation period of red foxes is also nearing its end. Gray foxes, which prefer to make their own burrows in heavily wooded areas, mated in late January and February just as the red foxes did, but their gestation period is a little longer (fifty to sixty days), so their young will not be born until April.

Gray foxes are even more elusive than red foxes, and we have had only a few glimpses of one over our years here. Except for Bruce. One early summer morning he went out to walk along First Field and, as he stood looking, a pair of young gray foxes came frolicking toward him. Oblivious to his presence, they played for several minutes before running back into the woods. Needless to say I was wildly jealous of his sighting and have yet to be as lucky.

The gray squirrels that I saw courting in January are having their first litters now. Their nests are tucked high up in natural tree cavities. The newborn are completely helpless, with eyes that remain closed for one week. At least seven weeks will pass before they can venture out of their nests. But by mid-May I will see the first fumbling emergence of young gray squirrels from the hole high in a massive dead oak tree at the edge of our woods. Day by day the young will grow bolder and more confident. Soon they will be chasing each other with as much aplomb as the adult squirrels.

Red squirrels, on the other hand, mate near the end of winter, and their young are born in April, May, and even June. They mature more slowly than gray squirrels, opening their eyes in a month and being weaned between nine and eleven weeks of age. For years they lived in our yard trees, keeping the gray squirrels in the woods, but more recently they have been entirely supplanted by gray squirrels. Seeing a red squirrel anywhere on our mountain is now a rare event.

Chipmunk males only begin searching for mates in early March, and their first litters do not arrive until April or early May. They, like the red squirrels, take a month before their eyes open and are weaned anywhere between five to seven weeks of age. Their mothers breed a second time in late summer, and by fall the woods are again filled with chasing chipmunks. Each is setting up and defending its individual turf—the area around its burrow—from other chipmunks.

Some cottontail rabbit females have already given birth. They breed in late January and February, and rabbits only need thirty days in the womb. All during spring and summer they continue to breed like, well, rabbits.

Opossums need even less time—thirteen days—but their young emerge in a totally helpless, embryonic state. Born in mid-March, the young opossums crawl hand-over-hand from the birth canal up to their mother's pouch, where they each grab a teat. But not all are successful in the climb, and since a mother opossum sometimes gives birth to as many as eighteen young and only has thirteen teats, some quickly perish. Those that survive remain clinging to their own personal teat, safe in the mother's pouch for fifty to sixty-five days, before they are able to emerge, clamber over their mother, and ride about on her back—the only time, incidentally, when most people will find them endearing.

Opossums, like skunks and porcupines, have a bad reputation. They are usually described as ugly, repulsive animals whose only defense is a cowardly one, that of playing dead by rolling over on their sides and setting their mouths into horrible, toothy grins. But although they may be animals of "little brain," like Pooh, they have managed not only to survive humanity's opprobrium but to thrive, extending their range much further north than their warm climate metabolisms should seem to permit. Here in the north you can always tell a brand new, first season opossum from an old one by looking at its naked ears. The old opossums, which have survived a winter or two, will have ears disfigured by frostbite.

I was reminded of mammalian reproduction when I noticed signs of activity in and out of the woodchuck den beside the Far Field Road this morning. It is comforting to know that no matter what the weather may be above ground, beneath my feet in dens and in tree cavities high above my head, the spring rituals of courting, mating, and birthing continue.

MARCH 19. Can we be having still another snowstorm? I groaned when we awoke to falling snow this morning. Bruce, though, appreciates March snows. He knows that with longer days of warming sun, the snow will quickly melt. So instead of facing each snowfall with dread—will it or won't it be deep enough to seal the hollow shut?—he whistles merrily. Another bonus to March snows, he points out, is that he can actually see them. In the darkest days of winter, he leaves the mountain before sunrise and returns after dark; snow and ice are a menace viewed only in the glow of the car's headlights.

Today, however, he can admire the hemlocks in the hollow bowed low with new snow, he can stop to watch the herd of deer filing silently down toward the stream for their sunset drink, or admire the ribbon of water flowing valley-ward over and around ice-polished rocks.

I, of course, have such visions of snow beauty throughout the winter months. Today I stood at the windows as the dawn light crept slowly over a black-and-white panorama of falling snow and watched a doe, covered with a thick coat of flakes, browse on a laurel branch burdened with its own layer of snow. Birds I had not seen since the Christmas cold flocked to the feeder.

But those birds were singing even as the snow sifted down over their perches. "Poor Sam Peabody, Peabody, Peabody," a migrating white-throated sparrow caroled from the grape tangle. "Hip-hip-hurrah, boys. Spring is here," answered a song sparrow. "Cheer, cheer, cheer," chirruped a cardinal from the top of a black walnut tree. "Peter, peter, peter," a tufted titmouse responded. Such music in the midst of cold and snow is not heard during the depths of winter.

Once I left our yard for my morning walk the woods were silent except for the sound I made as each foot sank down through the several inches of powder snow atop a crunch of sleet ice. Climbing up what is usually a gentle incline suddenly became a monumental athletic feat complete with laboring lungs and sweating brow.

Off came my knit beret and mittens. Down went the zipper of my ski jacket. Even when I reached the level road to the Far Field, I was warm enough to keep the hat and mittens off and the jacket opened. At times the snow fell so thickly that I could see less than three hundred yards ahead, and my glasses quickly clogged with fat snowflakes.

Where was spring in all of this? Although it officially arrives tomorrow evening at 11:30 when the path of the sun following the line of the ecliptic northward crosses the earth's equator, we may not see the true spring of warmth and sunshine and flowers for several more weeks. As Professor Henry Van Dyke once put it in his *Fisherman's Luck,* "The first day of spring is one thing, and the first spring day another. The difference between them is sometimes as great as a month." I hope he is wrong.

MARCH 20. Into the feeder this morning came the first field sparrow of the season, his pink bill his badge of identity until, that is, he begins to trill his plaintive song. That time will come in a few weeks when once again the mountain fields and edges will echo with field sparrow song. Then it seems as if every possible breeding territory in First Field is filled to overflowing with field sparrows. Yet there is still room, in late April and early May, for several indigo buntings, American goldfinches, and common yellowthroats to set up their own territories in our shrubby field environment. For almost a month and a half, however, these sparrows will have the field to themselves, and once they start to sing nothing, not even new arrivals, will deter them. To me there is no song quite so haunting as a field sparrow's melancholy tribute which continues even through the dog days of July and August.

MARCH 21. Most of the snow has melted again, and on this official first day of spring I spotted the first woodchuck out and about his burrow on the warm First Field slope just below Sapsucker Ridge. Birds still crowded the feeder though, and there are the usual signs that of all the months of the year, March is the hungriest time for the birds and animals that have been abroad all winter. Food is uppermost in all their minds. Mountain laurel leaves, buds, and twigs have become the survival food for deer, along with the tender, new, wild onion shoots that have sprung up at the top of First Field.

I took a midday walk to the Far Field thicket and discovered that large areas had been torn up by deer in search of food. I wondered how many deer it had taken to rototill so much earth and to liberally fertilize it with their pinto-bean-shaped scat. In some places the roots of trees were exposed and moss ripped out of the ground. It looked as if the deer were desperate for food, yet it has been an easy winter—so easy, in fact, that herds of twenty or more range across the top of First Field every evening, and deer numbers are higher than they have ever been on the mountain.

MARCH 22. This morning there was a hard frost but it quickly warmed up. I was greeted by a dawn chorus composed equally of robin, cardinal, titmouse, bluebird, mourning dove, and song sparrow, all set to the beat of a drumming downy woodpecker. Even the phoebe joined in for the first time.

As I neared the Far Field on a walk, I heard a loud gobble, followed closely by a second one. I pulled out my Lynch's Foolproof Turkey Call and whined an answer, but there was no response so I finally walked into the field. As I watched through my binoculars, a gobbler dashed across the Far Field and then slowed to a meander as he ascended Sapsucker Ridge. He had a beard reaching halfway to the ground and a bright, reddish purple head and wattles.

Later I spotted a pair of bluebirds on the wire behind the barn and a pair of American kestrels on the power pole where once they raised a family. Chipmunks "cucked," gray squirrels chased, and overhead, in a cloudless sky, a raven croaked as the thermometer rose to seventy-two degrees.

MARCH 23. Something in me does not like a wind. When the winds of March shriek over the mountain, as they did today, I head for the hidden hollows—little oases of calm and silence in the midst of a world seemingly convulsed by demonic forces. There I find the animals and birds that have also fled the wind, and for a while I can sit, enveloped by the peace of my quiet retreat.

Sooner or later I must rouse myself from my reverie and dash back through the raging wind. I try to ignore the ominous creaks and groans of every tree which could, I am certain, topple over on me without a moment's warning. "Who has seen the wind?" I chant over and over to myself as if the magic of that old children's poem will keep the wind at bay.

Years ago, when we thought we could find a place on earth which was still remote and untouched, Bruce began talking about the Falkland Islands. He hauled out maps and located old magazine articles that showed appealing pictures of five species of penguins—and little else. Like the ornithologist-naturalist Roger Tory Peterson, who is nicknamed "King Penguin," I have always been intrigued by the little-old-men appearance of those flightless birds, and I almost agreed to emigrate to the Falklands.

Then I read an ominous statement to the effect that hard winds blow over the islands almost every day of the year. There are also no trees on the Falklands, which seemed to negate any danger of uprooted giants crashing on my head, but I knew the constant sound and movement of wind, day after day, would quickly drive me mad. So long before the British-Argentine war, I rejected the islands as a sanctuary.

The only time I welcome wind is on a hot, sultry midsummer day. As a teenager living in southern New Jersey, I once attended a party on a muggy day so hot that even the basement "rec" room was breathless. Then the host put on a record guaranteed to cool us off. It was Vaughan Williams's *Sinfonia Antarctica* which he had written to document Sir Robert Scott's failed exploration of that icy continent. In the first movement, Williams used a wordless female chorus and a wind machine to depict the wailing wind. As I listened, my temperature began to drop and, ever after, when the heat of summer overwhelms me, I play that symphony.

But I never play it in March. The March winds remain a menace to me. Yet they are perfectly normal, as Hal Borland explains in his book, *Twelve Moons of the Year*. They are as "inevitable as the vernal equinox . . . the dying breath of winter, the first triumphant gasp of spring. It is the wind of change, the voice of seasons in transition." So I must bear with the March winds. April is coming.

MARCH 24. Fifty degrees at dawn and the smell of spring is in the air. Two male cowbirds are at the top of the Carolina poplar tree, and one starts to sing, followed by a phoebe and then a song sparrow. The northern flicker is back and calling from the dead oak at the edge of the woods. Both bluebirds and cardinals carol.

I found the first wood frog calling at our little, six-foot-in-diameter pond at the base of First Field this morning. I sat along the rim, waiting for him to emerge from the muck. After twenty minutes he came up and watched me with his golden, unblinking, froggy eyes. Then he swam closer, calling once, and finally subsided into silence as he lay prone amidst the green algae in the breeze-riffled pond. Water striders and fat back swimmers also skimmed the pond's surface. As I sat there I occasionally heard the thin cries of killdeer from some distant, unseen place far overhead, but they never landed in the field. Later, walking back up the road, I was elated to see the first mourning cloak butterfly fluttering in front of me. All signs of advancing spring; all promises of more to come. Enough portents to keep me content no matter what setbacks still lie ahead.

MARCH 25. Steve called collect from State College at eight this morning to report that he had seen a migrating Louisiana waterthrush when he walked down the hollow road at dawn to catch a ride to the university. Bruce looked at me quizzically. What kind of kids have I raised to count the spotting of a Louisiana waterthrush worth a collect call, his look seemed to say. Luckily he does not really object because he knows that it was his choice to marry a naturalist.

He was not surprised when I rushed down the road to look for the bird. Actually I was more interested in hearing the ringing song of the first warbler of the season. Looking more like a member of the dull-colored thrush family than of the brightly ornamented warbler family, the Louisiana water-thrush has a brown back, a white throat, a white eyebrow stripe, brown streaks on its white underparts, and long pink legs like its look-alike cousins, the northern waterthrush and the ovenbird. Although I moved quietly, I neither saw nor heard the bird. Later I checked local bird records and discovered that the earliest date ever recorded for migrating Louisiana waterthrushes in central Pennsylvania was April 10. Apparently, the bird Steve saw was rushing the season.

MARCH 26. The Shawnee Indians of Ohio called March the "Month of Worms" because that was the time when earthworms left their communal burrows below frost line, where they had spent the winter curled up together in a huge ball, and headed up through the thawed soil to resume their business of eating, mating, and excavating their individual burrows.

Tonight was the first warm, moist night of spring, so Bruce and I went outside with a flashlight to search for earthworms in our lawn. Before we turned on the light, we stood still and listened to the soft, rustling sounds the earthworms made as they pulled bits of leaves and twigs into their burrows. All around us the earth murmured with earthworm activity, and when at last we switched on the flashlight, we were amazed by the numbers and varying lengths of the worms we saw. Everywhere we looked, long, slender bodies stretched over the lawn, but as soon as the light hit them, they boomeranged back into their burrows like taut rubber bands.

Earthworms cannot see or hear, but they are extremely sensitive to vibration, light, and cold. When they emerge from their burrows at night, they keep their tails anchored to the entrance so they won't get lost. At the same time, they move their heads around in search of food, which can be anything from dead leaves to pine needles. Since, like birds, they have gizzards to grind up their food, they also ingest minute pebbles. After absorbing the nutrients they need from food, they eject what are called castings. These castings are more valuable to the soil than the most expensive fertilizer on the market, because they are extraordinarily rich in all the minerals needed to produce good crops.

It was Charles Darwin, in *Formation of Vegetable Mould,* the last book he ever wrote, who made most of the discoveries about the value of earthworms. He kept several pots filled with worms and earth in his study because he "wished to learn how far they acted consciously and how much mental power they displayed." Together with his sons he spent years testing their food preferences and techniques of food gathering. In addition, a lady friend collected castings for him from a small area. The results of her work led Darwin to estimate that earthworms annually ejected 16.1 tons of dry earth per acre and that every ten years their castings added one inch to the earth's layer. He also declared that approximately 53,767 earthworms could be found in one good garden acre.

Today scientists have proved that his estimates were low. On excellent land, where the lady did her work, estimates are that they eject at least 22 tons per acre. Even in heavy clay pasture, such as we have here, earthworms churn up 11 tons of topsoil per acre in a year. Furthermore, as many as 1,500,000 worms can live in an acre of superior soil.

I'm willing to believe that estimate since I saw for myself just how many worms can be packed into a small space. Several autumns ago David and I planted an indoor window box with lettuce, and David put five earthworms into it. In January I planted more lettuce seeds and grew them on the windowsill until early April when I transplanted the seedlings into our cold frame. Afterward, I carelessly dumped the dirt out of the box and was astounded to see dozens of earthworms of all sizes emerging from the soil. Curious to discover how many worms that bit of earth actually contained, I carefully sifted each lump of soil through my fingers and counted thirty-seven. A few of them were long and fat, others were medium-sized, and quite a number were very small. I decided that, unknown to me, they had been breeding in my window box throughout the winter.

May is their normal mating month, and although all earthworms are both male and female, they do need to unite head to toe for one night to cross-fertilize. After four days, a cocoon forms below each worm's head, where eggs are deposited. Then the worm slips the cocoon off its head, closes its ends, and buries it in moist soil. Four weeks later, one to eight young worms emerge. In less than half a year, the young reach maturity. During lives that can be as long as fifteen years, one earthworm can produce 40,000 offspring.

"It may be doubted whether there are many other animals which have played so important a part in the history of the world, as have these lowly organized creatures," Darwin concluded. Today many scientists agree with him as more and more research on earthworms reveals how valuable, complex, and even intelligent they are. Without earthworms, it is safe to say, our earth would not be nearly as fertile or diverse.

MARCH 27. Today was the peak of wood frog courtship, one of the highlights of spring for me. I begin to anticipate it sometime in the depths of January. "In only a little over two months the wood frogs will be courting," I tell myself during yet another ice storm.

Ever since I spotted the solitary male the other day, I have been listening for what sounds like the faint quacking of ducks. Except the sound is not made by ducks but by male wood frogs, newly emerged from their hibernation in logs and stumps or beneath stones or old boards in our wooded ravine. In a rush they have made their way to our pond, and they call to entice the females to come to them.

As soon as I heard them I slipped on warm clothing and crept down through the dead field grasses, trying to sneak up on the frogs. I hid behind a small bluff beside the pond to listen, but as I slowly edged my head up over the bluff to look, those sharp, froggy eyes detected me and they dove down into the water where I could not see them.

So I stepped into the open and settled down at the edge of

the pond, with my rubbered feet in the muck, to wait. Twenty minutes passed as I listened to the song sparrows caroling from the tops of nearby weed heads. A male phoebe repeated his name with monotonous regularity from the black walnut tree beside our barn.

Then, in the intense, prehistoric silence that settled over the pond, the first amphibian head appeared, its eyes just above water level and turned purposefully in my direction. I sat ramrod still as head after head emerged. All the frogs focused their golden eyes on me with unblinking intensity. The numbers vary from year to year, but there are never less than forty; one extraordinary spring day one hundred wood frogs were crammed into that tiny pond! I remember watching them for such a long time that my head grew dizzy from the din of noise and movement.

Today it took about an hour of silent watching before they trusted me enough to slowly begin the show I was waiting for. First one bold male, then another, started to call. Tentatively, a few more began swimming about, their long legs trailing out behind. The females are always outnumbered, and when one at last came hopping boldly up to the pond's edge and jumped in, she was instantly mobbed by seventeen males. They dissolved into a mass of tumbling, pushing bodies, but eventually one male frog was victorious and the others left him to the business at hand.

He climbed onto her back, clasped her body with his forelimbs, and held on tightly—an embrace that scientists call amplexus. Usually they stay in amplexus for as long as a day. The male's tight grasp eventually stimulates the female to lay her eggs. One spring as I watched a mated pair, the female actually emitted hundreds of eggs from her oviducts into the water while the male simultaneously emitted milt, a sperm-containing fluid.

This day I saw only six couples in amplexus floating just below the surface of the water. As I waited, several frogs

jumped out of the pond, including a brilliant, pinkish red female who landed on the bank and lay on the dried grass as if
she were dead. She was the most beautiful female I had ever
seen, looking more like a tropical frog than a member of the
usually dun-colored wood frog family. Some females are
faintly rosy, but on this one, even the usually black patch behind her eye was a deep red—the patch that gives wood frogs
the dubious distinction of being known as "the frog with a
robber's mask."

Curious to see if I could get close to her I went over, knelt
down, and touched her. She didn't move, and for a moment I
was certain that she was dead, worn out from the attentions
of so many beaux. Finally I tried to pick her up but she
squirmed out of my hand. She took a couple of leaps to the
pond outlet and then, as if she thought better of entering that
mob scene again, she sat motionless. This time I picked her
up and carried her home, eager to show her to the family. But
no one was around, so I returned her to the pond. This time
she leaped into it but the males paid no attention to her. It
was then that I noticed the first clump of frogs' eggs which I
assumed she had laid; she was no longer of any interest to the
males.

Wood frog eggs—a deep chocolate brown on their upper
surfaces and whitish below—are protected by transparent
gelatin that the female produces at the same time she emits
the eggs. Her clump will mass together with the usual fifteen
to eighteen clumps laid by other females at one end of our
pond. Each clump contains at least one thousand eggs, so the
potential for little wood frogs is enormous. But it is only potential. Although the pond is a thick soup of tadpoles when
the eggs hatch several weeks later, long after the parents have
returned to their permanent woodland homes, predatory
water bugs, spring peepers, and salamanders, as well as their
own cannibalism, quickly cut the population down. If we
have a dry spring and the temporary ponds where wood frogs

prefer to lay their eggs dry up too soon, the entire population perishes. Our tiny field pond has never failed to fill with water in the spring, but the more typical woodland pond we had for several years in a depression on Sapsucker Ridge failed two years in a row, and now no more wood frogs visit it.

Rana sylvatica, "frog of the woods," has been aptly named. It is the only frog that prefers to live on the land, hunting amid the ferns and partridgeberries for worms, bugs, and snails. Yet all my acquaintance with wood frogs has been at the edge of their vernal ponds, witnessing what seems like a scene from a prehistoric age: the incredible fecundity of amphibian life long before humanity evolved. Nothing else in nature is quite so wonderful to me as watching the courtship of wood frogs on a brisk March day.

MARCH 28. Returning up the road after another session with the wood frogs, I found the first buttery yellow disks of coltsfoot blooming in the median strip and along the edges of the old corral where it has grown for years. A green immigrant from England where it was prized as cough medicine, coltsfoot is frequently mistaken for dandelion. But unlike the dandelion, it flowers before the appearance of the large leaves which resemble, the settlers thought, the feet of colts. Once again it seemed as if coltsfoot would be the first wildflower in bloom on our mountain.

Then Mark came back from an exploration of the slopes above the road halfway down the hollow to announce that he had found a clump of skunk cabbage. I was surprised because for some unknown reason we have never had any skunk cabbage either beside the stream or anywhere else on the mountain as far as I could tell.

Mark offered to guide me to the spot, so I panted my way up the slope to the one plant he had found—two flowers and eight emerging leaves. The ground beneath the leaf duff was damp despite dry weather, providing suitable growing conditions for skunk cabbage, but it seemed unusual to find only one plant. It normally grows in patches.

I put my forefinger inside one of the sheathlike, purplish brown spathes which protect the flowering structure, called the spadix, to see if I could detect its heat. But because the day was unseasonably warm (eighty degrees), I noticed no difference between the outside temperature and that of the spadix. Roger Knutson, who has been studying eastern skunk cabbage for several years, has discovered that it can melt its way through early snow and ice because its flower maintains a constant seventy-two degrees (Fahrenheit) as long as the outside temperature does not dip below thirty-two degrees for more than twenty-four hours. When that happens, the individual flowers usually die, but new spathes keep emerging until the weather allows some of them to survive. Then they are cross-pollinated by beetles attracted both by the faintly sweetish smell of the flowers and the warmth inside the heavily insulative spathes.

The eastern skunk cabbage (*Symplocarpus foetidus*) is closely related to jack-in-the-pulpit, arrow arum, and green dragon, all of which have the spathe and spadix structure of the Araceae family. This family originated in the tropics, and while most of them—philodendrons, caladiums, callas, arums, and dieffenbachias—have stayed there, eight aroid genera moved north into the temperate zone. Each evolved special mechanisms for surviving in a cold climate. In the case of skunk cabbage, it has taken the heat-making ability of all the Araceae, which is normally used to make the spadix produce more odor to attract pollinators, and has used it instead to give itself a head start during the short northern growing season. All other Araceae, including those in the temperate zone, heat up for only a few hours or at most for parts of several consecutive days. Skunk cabbage stays hot for two weeks or more. Furthermore, the colder it is, the more heat it produces. In fact, its complex heating system is more mammalian than plantlike, "more like a skunk than a cabbage," as Knutson says.

Since its dark, heavy, marble-sized seeds take four months to develop, it is imperative that skunk cabbage flower early. Those seeds are sometimes stored by squirrels and other rodents, which accounts for its local dispersal and is probably the way our single plant arrived in the hollow. But from the time the seeds germinate, it takes between five and seven years before they are large enough to flower. So perhaps Mark's plant was making its debut this spring. I can only hope that its seeds will germinate around that parent plant so in years to come we will finally have a proper skunk cabbage patch on the mountain.

MARCH 29. All my watching has finally paid off. At last I have seen a red fox at the Far Field den again. About 9:30 this morning, as I was slipping down the usual far side of the field screened by the black locust trees, a large red fox emerged from one of the holes and looked directly across at me. At first I thought it had seen me, but it looked away. Then it trotted down to the far end of the field and did some nosing around with a few short pounces.

I took the first opportunity to ease myself to the ground and watched as it slowly trotted up the field toward me, veering, at the last moment, around the other side of the locust grove. It stopped once to sit and scratch. Then it proceeded up to the top of the field, headed toward a west-facing hollow, and passed within a couple yards of where I was sitting. At the top of the rise it paused once again, made a sudden pounce in which it turned a partial somersault, just as our cat used to do while hunting, and finally went on over the hill and out of sight.

Such a sighting pulled me back to the Far Field in the afternoon, hopeful that I would once again find a fox family in residence. To my unutterable delight, it was so. The vixen emerged from the upper den hole followed by three charcoal gray kits, all of which kept trying to nurse, just as I arrived at my usual watching spot behind the locusts. She finally shook off her kits and moved to one side above the hole where she sat and scratched and looked around.

A couple of kits stayed outside for only a few seconds and then went back into the den. Several minutes later the dog fox came bounding out of the lower den entrance and padded over to the scratching, resting female. They touched noses a couple times, after which he restlessly paced around and looked warily up the hill behind the den. He then tried to settle down on the ground but leaped nervously to his feet once again and peered up the hill. The female finally went back into the den accompanied by all the kits, and he joined them a few seconds later.

Judging from the size of the kits, they had been born early in March, several weeks earlier than researchers claimed for our region of the country. But then, despite the ubiquitous distribution of red foxes throughout four continents, surprisingly little scientific study has been done on them because of the difficulty in observing such wary creatures. J. David Henry followed hunting foxes on foot for fourteen years in Saskatchewan's Prince Albert National Park, where foxes have lost most of their fear of humanity; and David Macdonald conducted radiotelemetry studies of urban, suburban, and rural foxes in and around Oxford, England. Together they prove that red foxes are the ultimate opportunists, suiting their lifestyle to their environment.

It looks as if this year's litter is well ahead of last year's, and I would judge that they are already three to four weeks old. To be blessed with two years in a row of fox family watching is a privilege I will try to be worthy of, which means that every day my walk will take me to the Far Field. Suddenly my spring has taken on a special aura of delight that helps to quell the pangs I so often get when I think of the swift passage of time.

How many more springs will be allotted to me? Never

enough of them even if I were granted a hundred more. That is why I am less and less inclined to do anything else in spring other than to observe its passage as the years advance, knowing how limited my time on earth is and how much faster spring seems to go every year.

MARCH 30. This warm day has brought out what I label, for want of a better name, the "calling spiders." They are everywhere in the leaf duff along Laurel Ridge Trail, a phenomenon that I noticed for the first time only last spring. Could it be that this has gone on every spring and that I never noticed it until then, and that now that I am aware of it, I hear them wherever I walk? Is it a case of having my eyes opened on still another aspect of nature, or is it instead a sudden upsurge of the species which has brought them to my attention?

Whatever the reason, I cannot identify the spiders, but only describe them as having beige legs nearly an inch long, and two black lines running from cephalothorax to abdomen on their backs. They vibrate their two, small, black, curled-under papillae to make the buzzing noises I hear. I assume that it is some kind of a mating ritual but I have yet to see any consummation.

On the Far Field Trail there were at least four spiders in close proximity, all vibrating and ignoring each other until one came within seven inches of another, seemingly challenged it, and then rushed off. I sat on the trail for over an hour watching and listening to them. I finally decided that they were all one sex calling to the other, since none responded to the other. Each would pause after vibrating, take five steps, vibrate and pause again, looking as if it were listening for a response from underneath the leaf duff.

Last spring I had watched one vibrating spider running, calling, and then pausing until a silent, look-alike spider dashed up to within a yard of it and disappeared under the leaf duff, quickly followed by the first spider. Today there were no responders, and I finally abandoned them to their calling to search the spider books I own for answers. But so far I have found none. Spider behavior, it seems, has not been as intensively studied as insect behavior. With spiders I am always left guessing.

MARCH 31. On this overcast, dull day I sat near a ruffed grouse drumming log on Sapsucker Ridge. Suddenly I noticed movement around the base of a tree about six feet away. I tried to train my binoculars on the darting, gray brown creatures, but they moved too fast. I knew they had to be shrews with their small bodies (less than two inches long, with tails stretching another inch and a half) and their rapid movements in and out of the leaf duff, but I was not certain of their species.

I watched, mesmerized by their activity. Like animated jack-in-the-boxes they zipped a few inches up and down tree trunks, popped in and out of a rotted tree stump, and scampered behind tree roots, reappearing a few seconds later. It was impossible to count them but there was no doubt that what biologists consider to be extremely antisocial and bloodthirsty creatures, known to sometimes kill and eat each other, were behaving in a very sociable way.

Shrews are almost blind, depending on sound and smell for their livelihood, so they probably never knew I was watching them. These particular shrews also had a definite pattern in their activity—ten minutes of action, five minutes of stillness—which continued for as long as I sat there. Finally I left, elated but puzzled by what I had seen.

Then, as I walked along Laurel Ridge Trail, I heard a rustling in the leaf duff. There, almost at my feet, were more shrews darting about. I watched with fascination while they ran up and down a fallen log and leaped from a small knothole in that same log, all the while making small twittering sounds. Immediately I sat down with my back against the log and within five minutes they were darting along next to my outstretched legs and squeezing themselves through a small gap between my back and the log. One shrew even slammed into my back with a jar that made me jump. To them I was just another stationary object in the forest.

There were at least a dozen of them, and usually when one appeared, it was closely followed by another. Since many species of shrews do begin mating in late March, I thought the chasing might be preliminary mating behavior. But all the books and articles I later consulted said nothing about what I had observed. I did decide that I had probably been watching masked shrews since that species is a dull gray-brown, like the ones I had seen, and their masks are not noticeable in the field. Furthermore, they are the second most common shrews in Pennsylvania after the short-tailed shrew, and they prefer a woodland environment where they seek cover under fallen logs. Masked shrews also do occasionally venture out onto the ground surface, and their breeding activity reaches a peak in April and May. But I could find no description of that activity.

Shrews are elusive insectivores that spend their short lives burrowing through the leaf duff in search of earthworms, insects, and small vertebrates to fuel their incredibly high metabolisms. Most people mistake them for mice when their cats bring them home, but a closer look at the pointed snout, pinpoint-sized eyes, and hidden ears clearly distinguishes them from mice. Few scientists have the patience and expertise to study them, although some work has been done on the winter survival habits of the short-tailed shrew. No one, to my knowledge, has investigated their sex life. This, like the "calling spiders," seems ideal material for a budding biology graduate student to tackle. I, on the other hand, am content with the rare privilege I had today to witness a small part of the secretive lives of masked shrews.

APRIL

Andante

APRIL I. April arrived as April should, with intermittent showers and reasonable warmth. Two new birds also made brief appearances. The first, a hermit thrush with its distinctive reddish tail that dips slowly up and down whenever it perches, is both a spring and fall migrant. One winter I even saw a hermit thrush in the grapevines of Sapsucker Ridge in mid-January, and twice we have found one at the Far Field thicket during the Christmas Bird Count. But by and large it is a bird of passage, mute and secretive, only once sharing its song with me—the most beautiful song of all the thrush family.

The second new arrival this morning was a brown thrasher. Unlike the hermit thrush, it nearly always gives us several days of nonstop imitations from the black walnut tree in the back yard whether or not it stays to nest. During the first week of his return, the male is generally as reclusive as the hermit thrush, skulking about in the grape tangle or beneath the overhanging forsythia bushes. Then, in late April, he will begin to sing songs that are loud, assertive, repetitive, and sometimes imitative; the brown thrasher is a close relative of the catbird and mockingbird. In fact, I call him the "mocker of the mountain" in lieu of true mockingbirds which never nest here. Instead of a large, gray, noisy, all-night singer we have a sleek, cinnamon brown, long-tailed, all-day singer.

Once the female arrives and a decision is made to build a nest, no more is heard from the male. Furtively they build their nest, she lays three to four eggs which they take turns incubating for fourteen days, and after the eggs hatch they share in the work of feeding their nestlings such delicacies as insects, larvae, spiders, and worms. Back in 1912 Dr. Ira N. Gabrielson wrote in the *Wilson Bulletin* that a family he had observed in Iowa had consumed 247 grasshoppers, 425 mayflies, 237 moths, and 103 cutworms in twelve days. Here on the mountain we have watched them eat gypsy moth caterpillars and worms. Twice they obtained the latter by following after us as we dug up the garden soil.

Eight years ago we had an unusually gregarious brown thrasher family in residence. Once the nestlings fledged they grew bolder and bolder. I often watched them greeting each other in soft voices above the grape tangle or under the lilac bushes. At least that's what I thought they were doing. But a closer look changed my mind. In reality, the fledglings were begging for food from the male, while the female was incubating a second set of eggs.

The fledglings were almost as big as their father, but instead of having a cinnamon brown coat, they were a duller brown with a slight grayish tinge. The brown spots on their white breasts were brighter than the adult male's, and their eyes were distinctly gray rather than the bright reddish brown of mature thrashers. I was easily able to make the eye distinction because they were all attracted to the back porch steps where Mark had been cracking the previous fall's black walnut crop with a sledgehammer.

One afternoon, as I stood at the kitchen screen door, the male brown thrasher landed heavily on the steps, ready to treat himself to black walnut chips. He had a little competition from the resident chipmunk, but both settled down in relative harmony. Just as the bird was taking his first peck, two fledglings landed beside him, calling piteously and gaping their beaks open. Instinctively reacting, like Pavlov's dog, to the proper stimulus, he began feeding them and he never got even a taste for himself.

But the following day the scenario changed. Once again he sparred with the chipmunk. Then he began eating peacefully. As before, the two begging fledglings appeared. This time he literally turned his back on them and continued eating while they followed him so closely that they trod on his tail. They cried but did not gape open their beaks. He ignored them even when one rushed at him head-on, reminding me of a child stamping its feet in frustration. They did not receive a morsel of food for their trouble. Finally, first the male and then the fledglings flew off and I never saw them again. As abruptly as that the dependent became independent. For them the free lunch had ended.

Since then I have not had so intimate a view of brown thrasher domestic life. But every spring, when the male arrives, I hope that he will find himself a mate and set up housekeeping in our grape tangle once again. After all, according to ornithologists, if all our songbirds were entered into a singing competition, the brown thrasher would win the prize for variety because of his extraordinary ability to improvise new songs. Donald E. Kroodsma claims that it takes thousands of different songs to impress a female brown thrasher. Why this should be so has not yet been discovered. I only know that the brown thrasher's repertoire impresses me and makes my April days ring as he practices hour after hour in our back yard.

APRIL 2. The weak sunlight has brought out the pink and white trailing arbutus flowers along the Laurel Ridge Trail, and I knelt to sniff the first floral aroma of spring. Later I caught what smelled like a stronger version of the sweet odor of trailing arbutus, but I tracked it down to blossoming red maple trees. For years I have admired the beautiful red-and-gold flowers of red maple trees, finding them to be especially breathtaking when examined under my hand lens, but never before had I detected an odor from the blossoms. Further experimentation proved that only some of the blossoms on some of the trees were scented and that the scent was only detectable intermittently.

My last scent discovery of the day was along the stream. There the faint golden green blush of spicebushes in bloom caught my eye. They line our stream bank for half a mile down the hollow and are always the first shrubs to lighten up the woods. Best of all, though, is the spicy odor that emanates not only from the flowers but from the leaves, the twigs, and the berries. The latter were dried by settlers and used in place of allspice in baking.

APRIL 3. A wild sweet song from the tops of the tallest trees alerted me that the solitary vireos were passing through. Unlike most of the vireo family, which sport nondescript gray or olive garb, the solitary vireo has a striking blue-gray head with a white eye-ring atop an olive green back, a white breast, and distinctive white wing bars. Since they stay hidden in the treetops, I have learned to identify their high-pitched, melodious song. Furthermore, they have no competition from other singers in our woods this early in April, so it is easy to guess who the treetop singer is even if I have a difficult time, from spring to spring, recalling a song I only hear during the couple of weeks the solitary vireo stops in to visit en route to its breeding grounds.

APRIL 4. Another "dickey" sparrow joined the sparrow chorus this warm and beautiful morning. Chipping sparrows, with their agitated buzzing, are eager but unaccomplished "singers." Nonetheless, they sing and court and mate and raise their families quite openly. No skulking in the bushes for them. Their trick is to choose prickly, inaccessible nesting places such as the depths of the barberry hedge or juniper bushes. We have two juniper bushes, a long, old barberry hedge, as well as three smaller, red barberry bushes where I can always count on finding at least two chipping sparrow families. And whenever there is a commotion on the telephone wire it turns out to be a pair of chipping sparrows mating. Shame is not a word they are familiar with. Neither is discretion.

Along the Laurel Ridge Trail this morning I spotted the first spring azure butterfly flitting along like a bit of sky fallen from heaven. Tiny, delicate, blue-tinted creatures, they are the first of our native butterflies to emerge from overwintering pupae. Exactly the color of the sky, they are also called "blue" azures in spring. Later hatchings, in June, are much lighter in color, often fading into white. They lay their eggs in flower buds, and their larvae are particularly fond of dogwood although they also feed on black snakeroot, sumac, blueberry, and milkweed, all of which we have here in abundance.

APRIL 5. It was one of those perfect, blue-skied mornings we have been having this spring, and by 8:30 I was settled at the base of a large oak tree near the top of First Field, determined to call in a turkey with my Lynch's Foolproof Turkey Call. Slowly I made it whine like a hen, and within a few seconds I heard the far-off gobble of a male. Quickly I whined again. The gobbler answered immediately, only this time the sound was louder and much closer. Before I could compose myself, he came running down the woods trail a couple of yards to my right, followed by two hens. Then he veered sharply left, crashed through the laurel a few feet in front of me and burst out into the field. Coming to a halt less than thirty feet from where I was hidden, he fanned his tail feathers and slowly surveyed the area while the hens pecked calmly at the ground.

Cautiously, I touched that foolproof call. He gobbled back. Then he and his harem ambled through a patch of blackberry canes and into a nearby locust grove. Again I used the call. Not only did he answer but a second, smaller male also came running across the field from the opposite ridge. That turkey call was more than foolproof. It was magic!

Expecting a battle between the two gobblers, I saw instead a magnificent display by the first. He moved in slow motion, as if overwhelmed by the majesty of his bearing. His tail feathers were fanned to their fullest as he strutted under the locusts.

The second male, apparently acknowledging his superior, collapsed his own smaller fan, but he did remain near one of the hens. The other hen stayed with the larger gobbler who, however, seemed to be more interested in my, or rather Lynch's, calls than in her: every time I used it he spread his tail feathers. This bizarre exchange continued for half an hour. In fact, he was so intent in responding to me that the smaller male succeeded in spiriting away a hen. "My" tom never noticed. He kept looking for the shy hen who talked to him but did not join his harem as normal hens do.

The loyal hen continued eating as she moved slowly along, eventually wandering down the field and disappearing into some dried weeds. That brought the tom to his senses and he hurried after her, stumbling a bit in his haste. No doubt he had finally decided a bird in the hand was worth two in the bush.

I remained still for many more minutes, not wishing to break the spell that had settled over me. But gradually I returned to the world around me and noticed that a ruffed grouse was drumming on the ridge. Then two chickadees landed a foot away from me and didn't seem at all alarmed when I talked softly to them. They even remained to listen.

At last I rose to continue my walk through a woods luminous with light that lit up the beige forest floor and set every laurel leaf shimmering. It continued to be an enchanted day—I heard several more turkeys gobbling, and one ran across the path in front of me. My mind was so filled with turkeys that other creatures, like a hermit thrush, were only noted in passing. After years of watching turkeys run from me, I had finally succeeded in having them run (almost) to me.

APRIL 6. I never consciously think, "Today I will go out and look for yellow-bellied sapsuckers," and yet I always see them, one fine day in early April, quietly tapping and tippling in our woods. Like people who overindulge in liquor, the migrating sapsuckers seem oblivious to my presence, so deeply are they sunk into their cups. I have had several chances to watch them closely.

Usually I spot the males with their flaming red throats. The females, who migrate a week behind the males, have white throats. Otherwise both sexes have red crowns; dull, golden breasts and bellies; and black faces, wings, and backs accented by white patterning—two horizontal stripes on the face, a broad patch on each wing, and stippling on the backs. Altogether they are handsome birds and distinctive enough to be instantly distinguished from our five resident woodpecker species (pileated, downy, hairy, northern flicker, and red-bellied).

It is their sapsucking, for which they are nicknamed "sap-sipper" and "sup-sap," that sets them apart from other woodpeckers. Unlike their close relatives, insects do not form the center of their diets but are merely dessert, along with occasional wild fruits and nuts. Their major food is sap, which they obtain by drilling horizontal rows of small holes through the tree bark. In their southern wintering areas they are pariahs to the lumbering industry which claims that they not only disfigure the trees but make them more susceptible to harmful insects and disease and render the wood unusable for decorative purposes. Back in 1911, in fact, Waldo Lee McAtee concluded that "sapsuckers do not prey upon any especially destructive insects and do comparatively little to offset the damage they inflict. Hence the yellow-bellied sapsucker . . . must be included in the class of injurious species."

Since then, other researchers have found that sapsuckers do eat some injurious insects, namely the larch sawfly, the moths of forest tent caterpillars, spruce budworms, and beetles, as well as scale insects and other bark and tree insects. They add that controlling sapsuckers would cost more than the lumber industry's losses.

People in areas with resident sapsuckers continue to be ambivalent about the worth of the birds, especially when they drill into favorite yard trees. Those of us who only see them during migration are pleased to welcome them as visitors. To watch them at work on an alpine blue, April day, is one of April's joys, and I never feel the month is complete until I have found at least one sapsucker.

One spring day I discovered two only because I heard their peculiar mewing cries. When I finally located them, the two males were flying at each other around a medium-sized hickory tree, fighting over the possession of a favorite tippling tree already scarred with old sapsucker holes. I settled down to watch as one quickly routed the other and started to drink. He collected sap by bracing his tail against the tree at a forty-five-degree angle, gripping the bark with his feet, and dipping his beak into each hole two or three times. Each time he withdrew I could see his bill glistening with sap. Twice he had to stop and defend the tree from the other sapsucker, who tried his best to usurp the victor. Finally both birds flew off, and I went closer to examine the tree. I discovered a series of old holes that ringed the trunk almost to the ground. They had been plugged up with new wood and had not been reopened.

Although sapsuckers will often return to favorite trees every year and reopen old wounds, that sapsucker had made a new set of holes.

This morning I had an even longer and closer look at a yellow-bellied sapsucker. I was walking quietly along the Far Field Road when I heard a slight noise to my left. As I turned my head a male took off at eye level from a large sugar maple six feet away and flew about twenty-five feet onto another tree. Apparently that tree was not as attractive as the first. Even though I remained rooted to the spot, he returned, gave me what seemed like a penetrating look, and settled back down to drinking and drilling.

The tree's roots grow down along the bank, elevating it a couple of feet above the trail, and I noticed rows of old and new sapsucker holes in three separate series, beginning at the base of the tree. The flowing sap there was also feeding flies, but during the hour I watched the sapsucker, he never drank from any area but the third series of holes. Nor did he catch any of the insects attracted to the sap, as many observers claim sapsuckers do. Only once did he vary his diet: he made a quick dive to the ground, plucked up an insect, and flew back to his original perch.

For a while I maintained my motionless stance, then gradually eased myself to the ground. The sapsucker continued to glance at me occasionally, but seemed far more interested in drilling two new holes to continue the third series. Although I was within six feet of the bird, I never heard a sound; other woodpeckers are noisy when they make holes. After about ten minutes, while the sapsucker alternately drilled new holes and sipped from previous ones, the new holes were large enough to yield sap. His manner of sipping never changed. He would turn his head sideways and dip his beak into each hold two or three times, being careful not to touch the sticky bark with any other part of his body except his feet and the tip of his tail. I did not glimpse the long, brush-tipped tongue with which sapsuckers actually lap up the sap. He also defecated a fine stream every three minutes by quickly lifting his tail away from the bark and squirting a good foot or so from the tree. Twice he did move around to the other side of the tree, but most of his time he spent on my, and the sunny, side of the trunk.

At last he flew off, so I took the opportunity to stretch my cramped limbs. As I moved, two deer, which had been approaching me twenty feet from the rear, snorted and ran off, surprising me as much as I surprised them. Then I examined the weeping tree more closely. The sap flowed like water, and when I tasted it I detected only a slight sweet flavor. I counted eight new, round holes, although sapsuckers also drill holes or "sap wells," as they are called, that are square.

We named one of our ridges for the abundance of yellow-bellied sapsuckers our son Steve saw there during our first year on the mountain, and it is always somewhere on Sapsucker Ridge that I see the April migrants each spring. As if watching a sapsucker close up was not excitement enough for the day, I spotted three wild turkeys down near the Far Field thicket as I sat at the field edge. One had a short beard; all gleamed in the sunlight. But this time, when I used my turkey call, I only succeeded in scaring them off, ample proof, if I needed it, that every turkey is an individualist and what is sauce for the goose is not necessarily sauce for the gander.

APRIL 7. My glimpses of the red fox family have continued during the last week, but only today did I have the opportunity for a long sit. When the three kits were tumbling about outside, I realized that two had turned the color of dried grass, signifying a stage the previous year's kits had not reached by the middle of April. The third was darker, with more black. The grass-colored ones wrestled and played together, so the parents entertained the lone dark kit. The adults moved constantly—scratching, playing, checking on the kits—but both looked disheveled because they were shedding their heavy winter coats. As usual my view was half obscured by the locust trees, but I was willing to forego a clear look for the continual privilege of watching them without their knowledge.

Later, as I returned along First Field Trail, I watched two red-tailed hawks soaring above in an aerial ballet interrupted by a third hawk. Eventually one landed on a treetop high on Sapsucker Ridge while the other, after routing the interloper, came floating back, touched the back of the seated hawk, and landed a couple feet below it. Then both flew off toward the valley.

APRIL 8. I went walking in the early afternoon, beguiled by the sunshine and warmth. Near the top of First Field I spotted a woodchuck exploring around outside its burrow. As I approached, it went down into its hole. Carefully I walked twelve feet behind the burrow entrance and sat down. Several years ago I had done the same thing at a woodchuck burrow in the Far Field. To my surprise, that creature had popped its head out of its hole after about ten minutes, but it had been a young woodchuck, probably not as canny as an old one would have been. Today I decided to repeat my experiment with what was clearly a large, adult woodchuck.

When nothing happened after about ten minutes, I pulled my notebook out of my pocket and began writing. Suddenly the woodchuck's head shot out of its hole. I froze in mid-sentence, waiting for the animal to see me. Only it didn't. At first it faced straight ahead, its back toward me. Then it slowly turned its head—first to the right, then to the left. Surely that one black, beady eye I could see so clearly would spot me, but if it did, it gave no notice. Instead it twitched its nose, searching for a scent, and it wriggled its small, round, black ears. The wind blew toward me, giving the woodchuck no olfactory hint of my presence, and I made no sound that might alert it. But what of its eyes? Couldn't it discern my figure sitting a mere twelve feet away?

Apparently it could not see me because it hoisted its body out of its hole and proceeded to eat grass while I watched. Still, up until then, it had only had a sideways look in my direction. It did seem a little disturbed, though, because after a few minutes it circled in front of its hole and faced me head on. I waited to be discovered as it kept sniffing and looking toward me. But it did not stand up on its haunches to investigate as woodchucks often do, and it did not whistle its warning call.

Although a ruffed grouse drummed in the distance and a crow cawed, I felt wrapped in a cocoon of expectant silence as I watched the woodchuck. I wondered if it might show some sign of fear and charge toward me with barred teeth as a woodchuck usually does when it is cornered by a dog. After a long pause, however, it merely slid back down into its burrow.

I remained still and within two minutes it poked only its head out of the ground, swiveling its neck around to peer in my direction. It hesitated for a few seconds and then retreated. It repeated this maneuver, but whether we could have continued the game indefinitely I don't know. By then my legs and feet were badly cramped and I was forced to get up and move about.

Nevertheless, I had reached the same conclusion about woodchucks that I had about deer. They don't see very well and depend more on scent, sound, and movement than on figure recognition. If the wind blows my scent away from them and I see them first, it is easy to have close encounters with those wild creatures and get some idea of their private lives. And that chance, to add to my store of knowledge about wildlife, is what takes me out no matter what the season or weather. There is always more to be learned about the behavioral traits of even the commonest animals.

APRIL 9. Today it rained ducks and loons, grebes and great blue herons. Strong northerly winds coupled with a bitterly cold rain brought them down in three counties and gave us a rare look at waterfowl en route to their breeding grounds. Unfortunately, during the night when many waterfowl migrate, they mistook the rain-soaked, glittering highways for ponds and landed there, causing their own deaths and several traffic accidents as drivers swerved to avoid them.

Here on our dry mountaintop we only hear and see waterfowl in migration. I may hear a loon call as it flies overhead in spring, or see an occasional mallard on the river at the base of the mountain. But today I peered down at a white-winged scoter, a ruddy duck, and three horned grebes that were floating just above the bridge over the Little Juniata River.

Shortly thereafter the freezing rain turned to snow, and Steve, who had gone off to the local reservoir where waterfowl had been resting for the last couple weeks, called to tell me of the incredible numbers and variety of waterfowl he had seen. One small bog had had fifty great blue herons crowded in it, and the reservoir had had more than a thousand waterfowl—sixteen species in all. Later Mark appeared with a headless horned grebe that he had found in our stream just below the forks, the first of that species we have ever recorded here. Undoubtedly, it had landed there and, unable to take off, had made half a morsel for an owl or fox.

Grebes are powerful swimmers and divers, but they use their feet, which are not webbed like a duck's, as a rudder when they are flying since they have practically no tails. These feet are set far back on their bodies, making them very awkward on land. They need a wide expanse of water to run and flap across before they can take off. That is why so many were helpless once they landed on roads, small puddles, and streams. Other waterfowl, such as loons and white-winged scoters, were similarly helpless, and large numbers perished today.

Usually we are only vaguely aware that our home lies directly on the main migration route of many eastern waterfowl between the Chesapeake Bay and their northern breeding grounds. But on this day, when they rained from the sky, we were vividly reminded of our critical location below this waterfowl interstate highway.

APRIL 10. Our road through the hollow has become a highway of romance for the local grouse population. Every time we drive up or down we run into at least one lovesick female wandering down the median strip. Even when we stop, she doesn't move aside unless we get out of the car and chase her off like any barnyard fowl.

As I drove up this evening with Bruce we were stopped by a displaying male ruffed grouse, his tail fanned open and his dark brown ruff standing out like an upright fur collar around a Cossack's neck. Slowly he paced up the hill, regally peering both to his right and left as we watched.

Then I spotted the dead body of a female on the side of the road. Since she was intact with no sign of predation, we wondered what had killed her. As we emerged from the car to look more closely, another male grouse took off from behind us, and we thought perhaps the two males and the dead female were the same threesome Bruce had spotted over a week ago along the road. At that time the males were trying to outdo each other with their displaying while the female looked on.

All this behavior seemed odd because researchers claim that a male ruffed grouse drums to declare his territory and that females go from drumming male to drumming male in search of a mate. To have two males in a single territory is unusual. But then, as the years pass and my observations continue, I realize that our grouse, at least, seem to be highly individualistic in their behavior.

Take the lovesick grouse that twice in three springs crashed through our neighbor's kitchen storm door as well as the outer door and both times emerged dazed but intact; or the one that crossed and recrossed the road a total of six times in front of the car one evening. That befuddled grouse was probably participating in what biologists call the "spring shuffle," during which they frequently expose themselves in open places, the males strutting along and the females looking dazed and lost.

But best of all was the chance to watch a male grouse drumming, a spectacle I have witnessed only one April in my life so far. On that day I walked silently over trails that had been soaked by hard rains the previous night. As I neared the edge of the Far Field thicket, I noticed movement in the brush to my left. A ruffed grouse was drumming on a log barely twenty feet from where I was standing. The large, moss-covered drumming log was partially obscured by brambles, so, foot by slow painful foot, I inched forward whenever he was occupied by drumming.

For years biologists debated over how the ruffed grouse produced the muffled, staccato drumroll which accelerates and then abruptly stops. Was it done by beating its wings together, beating against the drumming log, or beating the air? Motion pictures of the process by Arthur Allen of Cornell University provided the answer. The male beats his cupped wings against the air, creating a sound with a frequency of forty cycles per second. At the height of drumming season in mid-April, the male often arrives at his site before daybreak and struts up and down before beginning.

This male was obviously preoccupied. I managed to creep within ten feet of him after fifteen minutes, but I still did not have as clear a view as I would have wished of his drumming. Then, just as I was anticipating hours of uninterrupted watching, I spotted a large woodchuck climbing over a nearby fallen tree. It was headed straight for the ruffed grouse log. As I stood frozen in place, it climbed up on the log beside the grouse, which startled him. The grouse jumped down to the ground, followed by the woodchuck, and while the grouse wandered around calling "wik, wik," in obvious confusion, the woodchuck came plodding toward me. At less than five feet from where I was standing, the creature stopped, turned its head to the side, and started to sniff the air.

I didn't move a muscle, even when it reared up on its haunches and held its paws in what looked like a begging pose. I was still hoping that the grouse would resume his drumming once the woodchuck was gone. Eventually the chuck decided that some danger did loom directly ahead and, turning around, it blundered back past the grouse and down into the thicket. The grouse, in turn, wandered off in the same direction. After another silent vigil I heard it drumming farther down in the tangle.

Since then I have frequently found drumming logs and I have sometimes startled a grouse away from one, but until today my glimpses of courting grouse have been fleeting. It has taken seventeen years for me to actually see the ruff for which the ruffed grouse was named.

APRIL 11. The first sign that nature plays no favorites appeared today along the Far Field Road. It was there that I found the beginnings of eastern tent caterpillar nests in a couple of the wild cherry saplings that grow beside the trail. *Malacosoma americanum* are early risers. They survive the winter as fully developed caterpillars still inside their eggs, which are protected from the cold only by a thin layer of shellaclike froth attached to cherry and apple tree limbs. They are able to endure the cold by "supercooling," a process of depressing the liquids in caterpillars' bodies below their normal freezing points without the formation of crystals.

In addition, their bodies get rid of some excess water and produce an antifreeze agent called glycerol. Glycerol is very important because it not only depresses the freezing point of body fluids but it also lowers the supercooling point. Scientists have found that in the summer when caterpillars first develop in their eggs, their bodies contain only 1 percent glycerol. By January it is 35 percent and remains that high until April when the weather is warm again.

Despite the incredible effort needed to make tent caterpillars winter-hardy, the eggs must have three to five months of cold before they can hatch. And the caterpillars, curled in their eggs, pass the season in the deep sleep known as diapause that is triggered by light and time.

Eventually they are awakened by the warmth of spring. They begin eating by first digesting the large yolks which were enclosed in their digestive systems the summer before when they were just eggs. Once they have obtained nutrition and strength, they hatch. Since the female that laid the eggs enriched the first ones with the most reserves, those eggs produce the most active, adventuresome caterpillars. They are the first to hatch, and later they become the most active moths.

Once they start hatching, the construction of tents begins in sturdy crotches of apple, crabapple, or cherry trees. They are built at random by varying numbers of caterpillars who use them as protection from heat, cold, and predators.

The main business of their lives, however, is to eat and molt. Their favorite foods are the leaves of the trees they build their tents in. Every morning, after it warms up, they emerge from their tents and move up the branches to strip the leaves. Six times they get too big for their skins and molt before they finally stop eating and search for secluded places to spin their white cocoons, suffused with brilliant lemon yellow. Often they choose the eaves of houses and barns.

Luckily for the cherry and apple trees, tent caterpillars have many enemies. Before they grow abundant hairs, they are consumed by numerous ant species. Shield bugs, although only one-eighth as large as tent caterpillars, are capable of piercing the caterpillars with their needlelike stylets and injecting a tranquilizing toxin. Then they hold fast and suck the caterpillars dry. Tent caterpillars are relished by black and yellow-billed cuckoos and parasitized by two species of ichneumon wasps and five species of tachinid flies. Even spiders catch them in their webs. During wet, late springs, the caterpillars are often victims of deadly bacteria, viruses, and fungi. In addition, the cherry trees themselves sometimes carry a virus that kills the caterpillars.

So, like all of nature's creatures, tent caterpillars do not have an easy time of it. Because they are natives rather than exotics like the gypsy moth caterpillars, they are kept in check by native enemies that have evolved to eat or destroy them.

APRIL 12. A thundery April day, but drawn as I am to the foxes at the Far Field, I risked a walk there near midday. Just as I settled down, a foxy red kit emerged alone from the den and quickly scuttled back after a loud clap of thunder. But in a few minutes the three kits suddenly burst out of the den at a dead run and streamed up the hillside to greet a returning parent, leaping up to touch noses with it, wrestling with each other, and wriggling with the exuberance of puppyhood. Several minutes later they repeated the whole scene for the second returning parent. As I watched I could not help feeling that what I was witnessing was a joyful experience every bit as heartfelt as when our three young sons had welcomed their father home from work each evening. That, of course, is anthropomorphizing, still taboo in many scientific circles, but then, I am not a scientist but a self-proclaimed lover of foxes.

Exactly why I am so fond of foxes has never been clear to me. Perhaps it began years ago when we lived on a farm in central Maine and I met my first red fox reclining on a boulder in the middle of our field. It was not asleep, and as I walked past, it looked directly at me in that intent, catlike way foxes have. Up until then I had never had a close look at a wild animal, and what passed between us that day remained a haunting memory. I wanted to know more about red foxes. As efficient predators with exemplary family lives, they could have a lot to teach me about a wild creature's relationship with its environment.

Now, as I continued my observations in our mountaintop field almost twenty years later, both parents stretched out for a short rest while two of the kits played together and the third played either alone or with whichever parent it chose to approach. I presumed it was the same lone, dark kit I had observed five days earlier because it was still in transition from gray to beige. One of the other kits remained a darkening beige to its playmate's already red coat, proof that fox kits of the same age do not always change their coats at the same time.

Because of their color differences I can identify the individual kits and even wonder if the constant companions are males and the loner a female or vice versa. Or has sex anything to do with their choice of playmate? Despite reports from scientists that fox play is often rough, whenever I watch them their play seems gentle. Perhaps that is because there are only three of them to share the food their hard-working parents provide. They may have already eliminated other, weaker siblings while underground, as some researchers claim they do. As far as I can tell, the red one, which should not have changed from its beige coat for another four weeks according to the books, is the dominant or alpha kit, its beige companion is second in the hierarchy, and the lone pup occupies the lowest or omega position.

After fifteen minutes, one of the parents again headed off to hunt while the other remained to watch the kits. Rumbles of thunder increased and the parent glanced up at the sky. Breaking my usual rule of waiting to leave until the fox family went into the den, I hoped they were not looking my way as I rose and hurried home, barely making it before a big storm hit with drenching rains.

APRIL 13. As I walked along the Far Field Trail early this overcast morning I spotted something large and black moving just on the other side of the fallen tree near the Far Field entrance. A look through the binoculars gave me the answer.

It was a fully fanned tom turkey's tail. I quickly sat down, hoping he had not seen me, and scraped the turkey call, but I never heard a sound. After about fifteen minutes I assumed that I had scared him off.

Then, just as I neared the Far Field, I spotted movement and froze. The male was busy displaying for a female who pecked nonchalantly in the dirt. I had a good close look at his blue face and red wattles before he realized that I was standing there. Unfortunately, unlike most mammals, turkeys see well and don't react only to movement. Silently they ran down along the edge of the field.

Later, as I walked back along the road another turkey went flapping off near the confluence with Laurel Ridge Trail. This is definitely a turkey spring.

APRIL 14. I have begun to wonder if the foxes know that I am watching them. Today my arrival at the Far Field was heralded by the warning call of a crow and the fleeing of three deer. In addition, the wind was at my back, effectively wafting my scent, I would think, toward the foxes. But although an adult fox was sitting above the main den entrance and ran up the hill as I entered the grove, two kits continued playing at the mouth of the upper den entrance, while the other adult sat above it along with the third kit. Far from appearing agitated, the adult remained seated and glanced only casually around. Then it stood up to check more closely on the kit. Finally the parent appeared to relax totally. It lay down and turned its back to me.

If they are aware of my presence that may explain why, whenever I bring a friend to watch with me, we never see the foxes. It could be that while they recognize and accept me, they will not tolerate other observers they are not familiar with. Some researchers claim that foxes do see well and will accept a certain critical distance between an observer and themselves once they get used to that person. To be accepted by a creature as wonderful and canny as a red fox would be an exquisite privilege not often granted to a human.

APRIL 15. For years the best place to observe the ephemeral woodland wildflowers has been the bank along our road up the hollow. It continues to have the greatest diversity of wildflowers anywhere on our dry mountaintop, but this spring the bank along the Far Field Road has also begun to nurture wildflowers. One, early saxifrage, is new to the mountain and proof that life, both animate and inanimate, never remains static. Always there are changes in numbers and sometimes in species as well.

After the onslaught of gypsy moth caterpillars six years ago, many of the old shade trees died. New areas, including the Far Field Road bank, were opened up to stronger sunlight, thus encouraging the growth of more plants and shrubs.

The delicate early saxifrage has a rosette of leaves that supports a single hairy stem containing near its top several small, five-petaled white flowers. The other members of the Saxifrage family on our mountain, foamflower and miterwort, both grow along the stream, but early saxifrage prefers dry woods or rocky fields. Although its scientific name, *Saxifraga,* means "rock breaker" because it is usually found in the cracks and crevices of boulders and outcroppings, it has only occasional stones to contend with on the Far Field Road bank.

Two small flowering trees that grow on the mountain, the shadbush and flowering dogwood, have also bloomed more gloriously since the canopy has been thinned. Shadbush begins the spring parade of white flowering trees in the forests of the East and along the vanishing hedgerows. Right now, on nearly every hillside and mountain slope, the shadbush is blooming, a bright contrast to the pale, pastel tints of red maple blossoms and spicebush.

Shadbush, or "shadblow," was originally named by settlers in the tidewater areas of the East who noticed that it bloomed when the shad came up to spawn in the rivers and streams. Other common names for it are "Juneberry" and "service berry," in tribute to the palatable purplish berries found on the trees in June and July. Those berries are so popular with songbirds and small mammals that they are always consumed before I get a chance to taste one. I can only accept the word of Euell Gibbons, the wild food expert, that they are excellent in sauces, pies, and muffins and that they can be dried, canned, or frozen.

Nevertheless, it is for their beauty and early blooming that I appreciate them. They seem to thrive along our trails, their long, slender, five-petaled, white flowers with reddish bud scales blooming in delicate clusters along silvery gray branches.

APRIL 16. I heard the first rufous-sided towhee sing off in the woods several days ago, but it was only today that I caught my first glimpse of the handsome black-and-white male with his foxy red sides in the grape tangle. The males return from the south a few days ahead of the females and set the woods ringing with their loud "chewinks" and "drink-your-tea" song. Towhees, like fox sparrows, brown thrashers, and white-throated sparrows, scratch vigorously in the leaf duff in search of insects and vegetable matter. Often, when the leaves are dry and the wind is still, I will think that I hear a large creature in the woods, only to discover that it is just a foraging towhee.

This day of cold and intermittent snow and wind brought back another common spring and summer resident, the blue-gray gnatcatcher. It migrates from as far south as Guatemala and always announces its arrival with a loud "zhee" as it forages high in the yard and forest trees. Later I will hear its slightly melodious, thin, squeaky-wheezy song, uttered so quietly that it is easily overlooked.

Smaller than chickadees, blue-gray gnatcatchers resemble miniature mockingbirds, with white breasts and long black tails edged in white that usually are sticking almost straight up from their blue gray backs. They are full of nervous energy as they flutter from branch to branch in our black locust and walnut trees in search of insects. These tiny birds build exquisite, small nests, often of cinnamon fern fuzz and oak catkins with linings of lichens bound together by spider webs. Their nests can be found anywhere from a few feet off the ground to seventy feet high on horizontal branches, although sometimes they tuck them into the forks formed by upright branches as well. They seem to have no preferences in the tree species they use for nesting, although a pair nearly always builds high in a black locust tree near our front porch. And like the chipping sparrows, they are noisy in their courting and their nest-building, so the location of their nest is never a secret.

Once a nest blew down in a windstorm, and we had a close-up view of it. It was just as lovely as the books described. No doubt they had to find a new site, but because we confiscated that nest they were unable to tear it apart and utilize its materials to make a second one as they are known to do.

APRIL 17. Please spare me from laments about the declining deer herd. I simply will not believe them. It's obvious to me that deer can tell hunters from nonhunters and only show themselves when the latter are about. This spring they have decided that none of our family are hunters so they have surrounded the house. They eat the daylily shoots on the slope, drink from the stream, graze in the yard, and lie in the flat area just below our back porch.

I can almost set my watch by them. I only wish they were color-coded so I would know exactly who belongs to whom and how many are repeaters in the course of one day. Sometimes a doe and her two yearlings amble up the slope. Other times there are three yearlings and two does. Often lines of thirteen to sixteen deer file down out of the woods and fan across the flat area, the lawn, and the driveway.

No time seems sacred to them. At dawn they can be spotted above the frog pond still resting. Then one by one they rise and slowly begin feeding. An hour later, as I walk past the garage, eleven heads, belonging to deer that have been grazing—both inside and outside the garden fence—suddenly pop up. Usually they take their own leisurely time to move off.

Of course they stay alert as they feed, heads instantly being raised and ears pricked up as soon as they hear a noise, but if they see that it comes from the house they go right back to their grazing. I can almost see them shrugging their shoulders and I imagine them thinking, "Oh, it's just Them. They're harmless."

This afternoon three deer were eating on the slope beside the barn. Mark went down to the barnyard to shoot some baskets, and those deer merely looked up while he shouted at them. Then they resumed their eating. One even lay down to ruminate as Mark proceeded to thud away with the basketball. However, if the noise comes from the woods they are instantly off and running. And if we have guests or a strange car sits in the driveway, they stay hidden discreetly in the woods.

The deer can tell our car from other vehicles. When I see their ears perk up around 6:00 P.M. I know Bruce is driving up the hollow. Sure enough, a moment later he appears and the deer calmly watch him pass. But any other motor sounds from the hollow scatter the herd in disarray.

Around noontime there is a rush hour in the grape tangle. Deer file in like caribou—along the ditch, up the slope, and in among the grapevines. I start counting and reach thirteen easily. Mostly they eat and drink just as we do at lunchtime. But today a yearling played with its shadow in the duck pond, prancing through the water, bending its head until it could see its reflection and then jumping like a frisky colt.

At dusk they all line up at the top of First Field to graze on the newly emerged wild onions and grasses, and each evening the numbers increase. Tonight we counted forty deer strung out across the field.

APRIL 18. For several weeks I have paid only occasional visits to the pond to watch the progress of the wood frog eggs and to look for other pond inhabitants. Two red-spotted newts are living there and feasting on frogs' eggs. These attractive salamanders, with red spots on their dark, greenish brown skins, live in small bodies of water and prey on tadpoles and frogs' eggs.

Usually the surface of the pond has several water striders of varying sizes gliding over it and numerous water beetles swimming just beneath the surface. Slowly the frogs' eggs have been assuming a tadpole shape, and five days ago, three weeks after the eggs were laid, I noticed tails wriggling back and forth in the egg masses. The following day the eggs began hatching.

Today I went down to look for other minute forms of pond life and spent several hours scooping up samples of pond water that was thick with black tadpoles. But once I found a twig adhering to a leaf which turned out to be a caddisfly larva.

I also found a Gordian worm, or "living horsehair," that resembles a very thin, six-inch-long wire. It writhed in the water and when I pulled it out, it dried up and tied itself into knots—hence its name, which refers to the old Greek legend of the Gordian knot.

Good tadpole food thrives in the pond—mosquito larvae, copepods, water scavenger beetles—and the pond was green with nutritious algae. In just a few more weeks, enough time to turn the surviving tadpoles into small wood frogs, our temporary pond will become no more than a muddy hole until next spring when the cycle of pond life will begin anew.

APRIL 19. The dogwood is blooming along the Far Field Trail. At the furthest edge of the field, I found, standing out from all the other trees, a perfectly shaped dogwood tree, shining in white splendor against the pale green understory of emerging tree leaves. Dogwood always grows best along the edges of the forest or in clearings like that one; I have also discovered lovely specimens gracing the edge of the power line right-of-way. Those that grow deeper in the woods usually produce only a few "blossoms," a botanically incorrect term for the white bracts that surround the true, small, yellow-green flowers in the center.

Why the name "dogwood," I've often wondered? Although dogwoods can be found in other parts of the world, the flowering dogwood, *Cornus florida,* is native to the eastern and central United States. But as with many of our plants, the name seems to have originated in England with a species whose bark could be steeped in water and used as an astringent to bathe mangy dogs. Others claim that "dogwood" is a derivative of "daggerwood," meaning that it was a strong wood useful for skewering meat over an open fire.

The wood *is* strong, heavy, and fine-grained, prized for making shuttles, tool handles, wheel cogs, mauls, golf club heads, pulleys, sled runners, hayforks, and woodcut blocks. The Indians used its roots to produce a scarlet dye and its bark as a quinine substitute. Even the small branches were valuable: by splitting the ends the colonists made toothbrushes.

While colonists, Indians, and artisans appreciated its usefulness and we prize its springtime beauty, the birds and animals favor the dogwood in the fall. Wood ducks, ruffed grouse, ring-necked pheasants, wild turkeys, bluebirds, cardinals, and a whole host of other gamebirds, songbirds, and mammals relish the scarlet berries that often hang on throughout the winter months.

Donald Culross Peattie was as enamored of the dogwood as I am. He too knew the joy of finding a perfect specimen and wrote in his classic book *A Natural History of Trees,* "Stepping delicately out of the dark woods, the startling loveliness of Dogwood in bloom makes each tree seem a presence, calling forth an exclamation of praise, a moment of worship from our eyes."

APRIL 20. Along the Far Field Road this glorious morning I heard a rustling in the leaf mold. As I watched, the leaf mold rose and fell like a miniature active volcano. Carefully, I pulled up the leaves and a plump hairy-tailed mole showed itself, or rather, tried not to show itself by curling up when I exposed it to the light. I couldn't see its eyes, but its long, pink, questing nose kept sniffing. I quickly let it go and back it went, humping up the leaf mold. Occasionally I glimpsed its shovel-like paws and snout, its two most important tools in digging and food-searching. But when I moved, it stopped altogether until I walked off.

Although evidence of mole work can often be seen, it is rare to glimpse the animal doing the work. Only once before have I had a good view of a hairy-tailed mole, and the two star-nosed moles I have found here were dead. Moles are members of the same order as shrews, Insectivora, but are not nearly as visible on the mountain—probably because they are less common and more nocturnal, sometimes emerging at night to feed aboveground. It is then that they are killed by the owls, dogs, cats, foxes, and snakes that prey on them.

Hairy-tailed moles are solitary creatures except when they breed in early March. By now their four to five young have been born, following a gestation period of approximately four to six weeks. Within a month they will be out of the nest and voraciously pursuing insects, snails, spiders, earthworms, and other underground delicacies as well as their favorite food, the larvae of beetles.

People who like to have impeccable lawns malign all species of moles because of their propensity to build tunnels both for nesting and food hunting, creating unsightly mounds aboveground. Since we are not lawn worshipers, we do not mind the tunnels, or the mounds. We know that moles are useful creatures who eat harmful insects, and we are pleased to share our turf (so to speak) with them. The added excitement of seeing one in action is all that I ask in return.

APRIL 21. In April I am drawn by the ephemeral wildflowers that line the hollow road. Already the hepatica has bloomed, and this cool, breezy, partially cloudy day tempted me to check the progress of other wildflowers along the road. It was, I discovered, a violet day.

There are two general groups of violets, stemmed and stemless, and we have several species of both. Stemmed violets produce leaves from an extended, upright stem and are represented by the smooth yellow violets and the long-spurred violets which grow in a blowdown area along the stream. The latter, with their pale lavender flowers, are particularly showy when they spread, as ours have, up the hillside; smooth yellow violets are also attractive, especially the transplanted ones in my back yard garden.

Stemless violets have both leaves and flowers coming from the crown of the rootstock on what appear to be separate, short stems. These include the familiar common blue violet, which thrives in fields and woods, and two that grow along the stream and road banks deep in the hollow—the dainty northern white violet and the earliest blooming round-leaved yellow violet.

The flowers we admire, though, are not the principal producers of seed. Those flowers are pollinated in the ordinary way by bee flies, gnats, butterflies and, in the case of the common blue violets, by bumblebees. But later in the summer, long after those flowers have died, cleistogamous flowers appear. Cleistogamy means "the production . . . of small inconspicuous closed self-pollinating flowers additional to and often more fruitful than the showier type," according to Webster's *New Collegiate Dictionary*. In violets, the cleistogamous flowers produce the most seed.

One botanist, Robert E. Cook, studied how wild violets grow by tagging them each spring and then following their subsequent maturation. Spring after spring he returned to find that many of his subjects never reappeared. If weather conditions are wrong, if one white-tailed deer steps on a fragile blossom, if cottontail rabbits find a patch and eat the leaves in their earliest stage of development, the violet never flowers and seeds. Under ideal growing conditions, it takes approximately four years for a single plant to produce its first true flowers, according to Cook.

How does a violet grow? In the mother plant a male pollen grain penetrates the ovule; the resulting embryo obtains food from the mother plant until it forms its own dormant seed, which can live without water, light, or mineral soil. Eventually that seed ripens and falls into the leaf litter where it lies dormant through one or even two winters. If, during the spring, the right conditions of light and moisture occur, the embryo penetrates the seed casing with a root and two cotyledons or first leaves. During that season it produces two or three leaves, and by fall the nutrients have moved to the roots and rhizomes of the new plant. A dormant terminal bud is formed which may survive the winter. If it does, the plant produces five to six strong leaves in the second spring. The third year the cleistogamous flowers develop, and in its fourth year of actual growth, the familiar spring flowers will finally appear.

Violets are not merely ornamental. They are also nutritious both to wildlife and people. The leaves and flowers of the common blue violet, high in both vitamins A and C, appear regularly in our spring salads. The larvae of many fritillary butterflies feed on violets by night, and the seeds are relished by mourning doves, ruffed grouse, dark-eyed juncos, pine voles, and white-footed mice. Wild turkeys dig up and eat their tuberous roots.

So inanimate violets, like animate wildlife, live precarious existences. The wonder of it all is that so many of them survive and thrive. What, after all, would our springs be like without violets in the fields, the woods, and even the waste places? Violets, the old herbalists wrote, "specially comfort the heart." Here on our mountain they still do.

APRIL 22. The woods along the Far Field Road this beautiful morning are full of singing ruby-crowned kinglets. So is the Far Field thicket. Although the first one appeared in the garage forsythia back on April 12, today is definitely the height of their migration and the woods are filled with their continual rivulet of song. "Look-at-me, look-at-me, look-at-me," they call from every tree and shrub. But their song is a more subtle trilling that warbles all over the scale.

These tiny, warblerlike birds, related to blue-gray gnat-catchers and the old world warblers, seem rather dull-colored, greenish gray birds until the males erect the red patches on the tops of their heads that give them their names. They are also friendly, fearless little creatures, and one day I sat beneath a forsythia bush as a male preened its feathers just a foot from me. I have also watched two males chasing each other from limb to limb of a Carolina poplar tree, prominently displaying their ruby crowns. The naturalist John Burroughs, in his book *Far and Near,* aptly described male ruby-crowned kinglets as behaving "exactly as if they were comparing crowns and each extolling his own."

Ruby-crowned kinglets spend their winters as far south as Guatemala and they breed in the evergreen forests of northern New England and Canada, so my time for watching them is very brief. But while they are here they make their presence known in loud, ringing tones, welcome calls in the still relatively quiet woods.

APRIL 23. I discovered the first box turtle of the season on First Field Trail near a seep. It had just dug itself out from its winter refuge below the frost line. Earth still clung to its shell as it plodded along in search, no doubt, of its first meal since last fall. This one had the blazing red eyes of a male and hissed when I bent for a closer look.

Year after year I find box turtles in approximately the same areas. That is because, as Lucille Stickel discovered, box turtle males maintain a home range of 330 feet. The females are slightly more expansive—370 feet—and they will leave their home range in June to lay their eggs. But they quickly return and the males never leave at all. Box turtles are definitely homebodies. They are also peaceful creatures. Since they do not defend a territory, fighting is alien to their nature. Even during the spring mating season later in May the males are rarely aggressive to other males. All in all, box turtles are admirable creatures, and I must confess that I always bend down to say hello when I find one on the trail.

APRIL 24. I spotted my first barn swallow of the season this clear, windy day, swooping exuberantly over First Field. In a few days the sky over the field will be alive with these striking, blue-black birds with deeply forked tails, cinnamon buff breasts, and rusty red throats. They seem to be in continual communication, calling to each other as they seine insects from the air or lining up to chatter on the telephone wire.

Until we started raising chickens we had no barn swallows, but the first spring we bought our chickens a pair of them arrived to nest in the barn basement. At first I was puzzled as to why they should arrive with the chickens, but since they line their nests with either poultry feathers or horsehair, I am convinced that they were just waiting until we had the sense to buy chickens or horses.

Since then their population has increased year after year, and by August I can sometimes count as many as fifty coursing over the field. Most of them nest in the barn rafters, but some have chosen to use old phoebe nests at the top of the veranda columns for second nestings, so occasionally we have had swallows swooping into our veranda as well. I never realize how much I watch them until they leave in the third week of August, heading for their wintering areas as far south as southern Latin America.

APRIL 25. Today we had city guests who are interested in nature. They were eager to glimpse at least some of the creatures I claim live here, particularly the red foxes. Since I have not seen them for eleven days, and since our wild creatures almost never perform for guests, I was not too hopeful. Nevertheless, I played the game, walking swiftly ahead of them to scout out the Far Field and seeing absolutely nothing.

"Well, can we at least look at the fox den close up?" one of the guests asked. "I've never seen a fox den before."

I agreed reluctantly, worried that if the foxes *were* still there, our approaching close to the den would surely scare them away for good. On the other hand we did owe our guests something in the way of nature-watching. We talked in low voices as we neared the den just in case the foxes were around, but all we saw was an empty hole. It seemed strange to me to be so close to what I have been watching from a distance for so long. I felt as if I were trespassing.

Then, just as we were turning away, Steve said, "Look, there's a kit." Slowly we all turned back and there sat the foxy red, alpha kit at the den entrance watching us from fifty feet away. It seemed perfectly composed and remained still as we all admired it. I even spoke quietly to the lovely creature, something I often do to animals and birds. This seems to allay their fears or at least to delay their fleeing time.

But it was we and not the kit who finally broke away. Our guests were ecstatic and so was I. For once our wildlife had cooperated. We later saw innumerable singing solitary vireos and at the top of First Field, while we stood gazing at the view, an osprey, as if on cue, came soaring up along Sapsucker Ridge. Those guests went away convinced that we live in one of the nicest places on earth.

APRIL 26. The call from the fire department came after Bruce and I had gone to bed, and Steve answered it. Folks in town had been watching a fire on our ridge from the shopping center and through binoculars had determined that there were people around it. The fire company wanted to know if we were aware of the fire and since we were not, Steve and Mark offered to walk up and check on it.

Mark came running in a half hour later panting from the exertion. They had found a fire, obviously out of control, burning along the edge of an open rock slide. Whoever had started it had fled after trying to put it out. Steve had remained to try to keep it contained while Mark had raced home to call the fire company.

Fire on the mountain is never a joke, and tonight the howling wind made it a terror. Mark telephoned while Bruce dressed. Then Mark ran back to rejoin Steve far down Sapsucker Ridge, several hundred yards beyond the top edge of the First Field through the dark woods. It took the small fire truck, carrying four people with tanks of water on their backs, thirty-five minutes to make it up our mountain road. They drove as far as they could to the corner of the field and then, led by Bruce and a powerful flashlight, went trekking into the woods. By the time they reached Mark and Steve the fire had died down, but they did use their 160 pounds of water to dowse it as best they could. Then they covered it with rocks.

By 11:30 the excitement was over. But I could not help remembering the frozen February night a little over a year ago when we had a chimney fire. That time two trucks laden with a couple dozen men and one woman struggled up our narrow, icy road and spent all night trying to contain the fire with the least amount of damage to our house. Part of my study and Mark's bedroom had been scorched, and the chimney had eventually disintegrated. We had lost little except our confidence that fire could never happen to us. Both times we have been lucky, but our respect for the dangers of fire have increased enormously. So has our respect for the local volunteer fire companies and the people who so generously gave up their sleep to help us.

APRIL 27. This has been a warbler week beginning two days ago with both a yellow-rumped and a black-throated green warbler along the Laurel Ridge Trail. Yesterday I saw a pair of male yellow-rumps flitting in the treetops of Laurel Ridge and heard the first ovenbird of the season. And today, despite the frigid weather, a black-and-white warbler appeared along the Far Field Trail. At times like this, when spring seems once more to be in retreat, the appearance of the first warblers brings hope. Warblers, after all, are tropical birds, and if they have deemed it time to return, it is bound to get hot.

APRIL 28. The gypsy moths first reached our mountaintop thirteen years ago, and every year since then their caterpillars have hatched on or about April 28. This year is no exception. At first they look so inoffensive—small black caterpillars emerging from the protection of the beige, powdery egg masses scattered along tree trunks. Each mass contains dozens of tiny, translucent, ball-shaped eggs, and so today each is alive with wriggling caterpillars.

But inoffensive as they look, they purposefully climb up into the treetops in search of succulent new leaves just emerging from their buds. Within a few days the newest of leaves will be shot through with holes. As the caterpillars grow larger and fatter on tree leaves, they continually molt their skins. Their voraciousness increases with their size so that during a year of heavy gypsy moth caterpillar infestation, the canopy is thinned to early spring sparseness. By then it will be late June, and the caterpillars will be descending the trees for the last time to pupate under tree bark or on rocks. By mid-July, first the small, dark, male moths will emerge and then the larger, white females, and after an orgy of mating, egg masses will be deposited on the trunks again.

It is a cycle once foreign to the New World because gypsy moths evolved in Europe along with their natural enemies. In 1868 an entomologist from Medford, Massachusetts, who hoped to make silk from them, imported the caterpillars from the Old World and a few of them escaped. Over the years they have expanded across much of the eastern United States, despite efforts to deter their advance with poisons and, more recently, with the importation of some of their natural enemies.

When the infestation reached its height here eight years ago we steadfastly refused to spray and watched as most of our trees were defoliated. Many of the old ones died but some, like the young Norway spruce trees we had planted at the top of First Field, grew another set of needles. The younger, healthier trees also survived even after a second, less severe defoliation the following year. Although disease and overpopulation killed most of the caterpillars the second year, the population has once again been increasing.

The experts say the damage will never be so severe as that first year because already the local bird and mammal population exploit them as a new food source, eating the eggs, caterpillars, and even the moths. For instance, black-capped chickadees eat their eggs in the winter, and black-billed and yellow-billed cuckoos, heavy predators on gypsy moth caterpillars, return every May to feast. Cedar waxwings also come keening in just as the caterpillars reach the delectable stage. But each spring there are more caterpillars, and I await with dread another spring when the trees will be mostly bare.

On the other hand the changes to the forest have not all been bad. Until gypsy moths killed some trees, red-bellied woodpeckers did not breed here. Now they not only breed but they also winter on the mountain. Other woodpecker species have also thrived. So have the cavity-nesting mammal species. And the opening of the canopy has led to a greater diversity of wildflowers and shrubs. As with all of nature's creatures, even those which do not naturally belong here, there is a gradual balancing between the predators and their prey. Over the years I will continue to watch as gypsy moth caterpillars are slowly absorbed into our ecosystem.

APRIL 29. "The wood thrushes always return on pea-planting day," I told David this afternoon as we crawled along the ground poking pea seeds into the warming earth. It was a day so perfect that being alive in the sunshine was all that anyone could wish for. I, for one, was peacefully content to plant peas and listen to the calls of all the other April birds. But still my ears remained finely tuned in an effort to pick up the first strains of wood thrush music.

David, however, was doubtful of my prescience. "How do you know for sure?" he persisted.

"Because they always have before," I answered more confidently than I felt. The first strains of wood thrush song had usually come in the evening when I was finishing up the planting but today, with David's help, the peas were planted by five in the afternoon. At that time no wood thrushes sang.

All through dinner I assured the family that wood thrushes would be singing at dusk. Everyone smiled dubiously at me. Then I stepped out on the veranda to listen to the evening bird chorus. A cacophony of song greeted my ears—our yard, field, and woods reverberated with the sound of bird music.

Suddenly I cocked my head toward Sapsucker Ridge. "Listen, do you hear him?" I asked David. He gave me an incredulous look.

"Wood thrush," he answered. "You were right." I could tell he was impressed.

I, on the other hand, was relieved. My reputation as soothsayer was intact. Of course I had no guarantee that the wood thrushes would return on pea-planting day. On the other hand my bird list shows that they have returned on or about April 28 ever since we moved here, and I based my prediction on that fact. Even so, while natural events always occur here in the same sequence, they do not always occur on identical dates from year to year. So I was sticking my neck out for effect and I knew it.

Wood thrushes, *Hylocichla mustelina,* have been aptly named, since *hylocichla* is Greek for "forest thrush" and *mustelina* means "weasel-like" in Latin, referring to their tawny-colored heads and backs. Although they are handsome birds with their brown spotted, white breasts, it is their song that haunts the souls of those of us who love wood thrushes. Until they return and fill our woods with music, spring has not truly arrived.

APRIL 30. I have not seen the foxes since the day we spotted the alpha kit at the den entrance, and today my worst fears seemed justified when a woodchuck emerged from the main fox den entrance while I watched. It looks as if the foxes have deserted the den and that it has been taken over by a woodchuck.

But it is not to be so simple. I revisited what I had always thought was an old woodchuck den at the edge of the Far Field, less than a hundred yards from the fox den. But outside the den lay the rotting remains of a dead woodchuck, its head, neck, and a pair of legs still intact. The hole itself appeared to be freshly dug under the roots of a tree but with only the one, well-tamped entrance. Below it lay a pile of wild turkey feathers. Have the foxes not only found a new home but have they violently removed its previous owner?

MAY

Crescendo

MAY 1. How wonderful to be born in May, the love-liest month of the year! After all the false starts and tentative green shoots of March and April, the first begin-nings of spring flowers and spring courtings, May erupts with blossoming flowers and trees, courting, singing and nest-building birds, and the emerging young of the season—tur-key poults, just-hatched grouse, wobbly-legged fawns. My world, which has been open and visible ever since the leaves fell last autumn, suddenly closes in with tropical greenery, and it becomes harder to identify the assorted rustlings I hear as I walk the trails. For in May I roam the woods from morning until evening, searching for rare and elusive wildflowers, hop-ing to make still more discoveries down in the hollow where I know new species lie hidden.

Today, clad in leaky, rubber-bottomed boots, I walked up-stream in search of new wildflowers, not, I might add, most people's idea of high adventure. But then adventure is where you find it, and it is often the little, home-brewed activities I decide on the spur of the moment that remain treasured memories long after I have forgotten more spectacular vaca-tions in exotic places.

The easiest part was the mile-and-a-half walk down our

mountain road during which I tabulated the wildflowers growing on either side of our dirt track. First came the fragile looking, white rue anemones that tremble in the slightest breeze. But although they may look fragile, they are the longest lived of any spring wildflower on the mountain, often holding their blooms for a month and a half. Above them on the bank were the just-opened, inch-long, orchid-colored blossoms of fringed polygala or gaywings, and nestled among the rue anemones were a few clumps of pale blue hepatica fading to white.

The leaves of both true and false Solomon's seal were unfurling, while those of wild sarsaparilla and yellow mandarin had just emerged from the ground. Common blue, northern white, and smooth yellow violets as well as wild geranium and foamflowers were in blossom. But loveliest of all were the many large beds of wake robins or purple trilliums *(Trillium erectum)* growing along the stream bank. Nicknamed "stinking Benjamin" because of their strong "wet dog" odor, they are chiefly pollinated by carrion or flesh flies, so they do not need a sweet odor to attract bees. Although the majority of their flowers are liver red, occasional blossoms are pink, salmon, greenish, or white. We have a number of yellowish white ones that mingle with the liver red flowers, making our display even lovelier.

After an hour of slow walking I reached the bottom of the road and the mouth of the stream. Then I clambered down over the steep bank and my real adventure began. There, in a flat, grassy area, I found the first new wildflower, garlic mustard *(Alliaria officinalis)*, an alien with an odor of garlic which grows along roadsides, waste places, and wood edges.

Next I plunged into a heavy growth of rhododendron and found myself climbing under as well as over its gnarled branches as I tried to keep to the narrow stream bank. Even so, I had to cross and re-cross the stream numerous times, sometimes crawling underneath fallen trees blocking the way or walking in the streambed itself when the bank disappeared. Despite wet feet and a sweaty face, I proceeded steadily upward through a grove of hemlocks that sheltered at their bases many more clumps of the striking white clintonia than I had ever known we had. Indian cucumber-root was also abundant, and I found the first and only jack-in-the-pulpit plant in the hollow area growing beside the stream.

My other new-to-the-hollow flower was mitrewort, which had previously been overlooked because it grows among the more abundant foamflowers or false mitrewort. However, viewed closely, mitrewort, with its single stalk of tiny, fringed, white flowers flanked by a pair of stalkless leaves, is very different from the naked-stemmed, white, feathery flowers of false mitrewort.

For over three hours I walked upstream, getting hungrier, hotter, and more tired by the minute. Finally, after a mile of struggle, I climbed back up the bank to the road, having seen the wildest, most inaccessible part of the streambed—having had, in fact, a small adventure that gave me an entirely new perspective on our mountain stream.

MAY 2. This morning the sky was blue, a gentle breeze rustled the emerging tree leaves, and the mountain glittered in the lucent air. I rejoiced to be alive and free to roam on such a day. Again I was in search of wildflowers, this time along the mountain top. But as I descended still another hill in my search, I suddenly spotted a furry white creature about sixty feet ahead of me in the new green growth of the Far Field.

I paused and looked, thinking it was a stray dog. But it didn't look like a dog. What other wild animal was white? An opossum maybe? Certainly not. It was too pudgy to be an opossum. What it looked like was a small, white, bear cub. Not possible, I thought, but I glanced nervously around for a sow just in case. Then the creature, which had been eating the grass, raised its head and I knew in an instant what it was. A white woodchuck. White woodchuck? Was there such a creature? Slowly I eased myself down to the ground and just as slowly I raised my binoculars to my eyes.

Heart beating fast with excitement, I looked it over. It had dark eyes, a black nose, and a little gray on its ears, but otherwise it was all white. Not so small either, I noted, marveling that such an oddity could have survived to maturity without camouflage. It was so busy eating that it never noticed me, and I wondered if I were watching it break its winter fast.

As I sat there the breeze continued to blow, but the warmth brought out flies that buzzed around my head and the woodchuck's. Occasionally a bird called from the nearby Far Field thicket and the animal would peer in the direction of the call. However, it did not seem as vigilant as most woodchucks are. In all the time I watched, I never saw it rear up on its hind legs to look around as woodchucks usually do when they are out in the open.

After a half hour of observing, I decided to move closer. Every time the animal's back was toward me, I slowly and quietly walked forward, finally halting to sit on a large log twenty feet away from it. At such close range, I was able to examine its eyes more closely for signs of albinism. While the eyes themselves were dark, the rims around them were pink, a sign of partial albinism.

I didn't sit particularly still anymore, but the woodchuck never saw me. It continued to tear off and consume large mouthfuls of the grass and weeds surrounding it, and as it fed it kept moving closer to my log until it was ten feet away from me. Why didn't it notice me? Surely I didn't blend in with the green background any better than a white woodchuck did?

Then suddenly the animal stopped feeding and walked over to the log. I sat transfixed and unbelieving as it climbed onto the log and started shuffling toward me. Couldn't it see me? Was it blind? Yet it didn't fumble as a blind animal would. When it was three feet from me I spoke up.

"Hey, fellow," I said quietly. "Don't you see me?" It

stopped in its tracks and for the first time looked directly at me. I made no other sound or movement, and it climbed off the log and ambled behind a tree where it continued to watch me. It stayed there until I got up and walked toward it. Then it slid slowly down into its den beneath a tree root at the edge of the Far Field thicket.

MAY 3. As I stepped out into the yard this morning I heard the loud "wheep, wheep," of the first returning great crested flycatcher. Throughout the day I heard them calling all over the mountain. This kingbird-sized flycatcher with rufous bright wings and tail, gray neck and breast, yellow belly, and a bushy crest which it raises when it is excited, is a favorite of mine—probably because its explosive call is so easy to identify. Although it is more often heard than seen, I have had a couple of close-up views over the years. The first bird I caught flying frantically around in our shed, and I was surprised, when I held it, to learn how small it really was. Certainly its size (eight to nine inches) does not match the strength of its call. The second was an unusually curious bird that took, one spring, to landing on a locust branch near our front porch and cocking its head this way and that, clearly examining whoever was sitting on the chaise longue.

Another noticeable bird in our dooryard this morning was the common yellowthroat flitting around in the forsythia hedge. Common yellowthroats also have an easy-to-identify song. "Witchedy, witchedy, witchedy," they call with monotonous regularity. But what striking birds these warblers are. The male has a bright yellow breast and throat and a large black mask that is unmistakable; the female is less distinguished with only a yellow neck and breast. Both birds have white bellies and olive brown backs, and they hold their long tails at a wrenlike tilt, especially when I approach their nests, hidden somewhere in the depths of the undergrowth in the power line right-of-way. No matter how much they scold and I search, I have never yet found a nest.

Along the Guesthouse Trail I spotted another warbler, the ovenbird. Not nearly as flashy as the yellowthroat, it looks more like two other dull-colored warblers, the northern and Louisiana waterthrushes, with its brown back, white breast heavily streaked with brown, and long pink legs. It is distinguished from the waterthrushes only by the orange stripe outlined in black at the top of its head and by its penetrating "teacher, teacher, teacher" call. Today's ovenbird plunked along beneath the laurel and stopped to regard me, cocking its head and fixing me with its quizzical eye. Ovenbirds nearly always have what seems to be an interested-in-you look when I catch them unawares.

The Far Field Road was also awash in warblers today, chiefly male American redstarts. In fact, during the spring and summer I think of the Far Field Road as "redstart alley." The black male—small and butterflylike—has bright orange wing and tail patches; the female is olive brown with bright yellow wing and tail patches. The area along our Far Field Road is obviously prime nesting territory, and these less than shy birds zip about the trees calling "teetsa, teetsa teet" from morning until night. Once they start nesting, though, they are as secretive as any other songbird. Despite great exertion on my part, I have not found any of their nests either.

While great crested flycatchers, common yellowthroats, ovenbirds, and American redstarts are all nesting species, I had a glimpse of an uncommon migrant as well along the Far Field Road. At first I thought that it was a black-and-white warbler but a closer look revealed its distinguishing white cheek and throat and grayish back. A male blackpoll warbler sat quietly on a tree limb as I adjusted my binoculars. Definitely a day for the birds, as so many days in May are wont to be.

MAY 4. I was awakened at dawn by the frantic-sounding calls of a whip-poor-will right below our bedroom window, and once again hope sprang eternal. Would we, after so many years, have another nesting whip-poor-will?

I clearly remember our last resident whip-poor-will back in 1977. Bruce's father was still alive then. In fact, it was his last spring. He had been so delighted to hear whip-poor-wills once again, a call he had not heard since his boyhood. And this resident was unbelievably bold. One evening in May we were nearly blasted out of the living room by his call. It sounded as if he had flown into the kitchen, but when Steve went sneaking around the back of the house, he found the bird on the back porch. It took off from the porch steps when it saw Steve, but it stayed around our home grounds all summer, calling with monotonous regularity every evening from 9:15 until 9:45 and sometimes also at dawn. We all grumbled then about losing sleep. Little did we know that he was to be our last whip-poor-will except for an occasional call during spring migration.

There is no doubt that there are fewer whip-poor-wills around now than there were years ago. Last spring I heard no calls at all. Unfortunately they are difficult to find and study, and researchers are having a hard time documenting their numbers. Why there has been a downturn is not yet clear, but three possibilities have been proposed—the growth of brushy farmland (one of the whip-poor-wills' preferred habitats) into mature woodlands, the increasing urbanization of the East, and finally, the effects of insecticides. Despite Rachel Carson's landmark book *Silent Spring,* we continue to use poisons with unknown side effects on our farmlands, with devastating consequences for insect-eating birds. Roger Tory Peterson has speculated that spraying for gypsy moth caterpillars has diminished the whip-poor-wills' food supply by destroying moth caterpillars in general, especially the large, showy luna and cecropia moths, both favorites of the whip-poor-will.

This morning I did not grumble when I was awakened by

Appalachian Spring

the whip-poor-will. I rejoiced. Oh, for more sleepless, whip-poor-will-filled dawns.

MAY 5. This day may have been damp and heavily overcast but I had my reward just the same. While I was still inside, I heard a turkey gobbling. Figuring he must be close, I went out to put on my boots, heard more gobbling, and stood up to try to locate its source. The sound came from First Field down beyond the shed.

Two tom turkeys were leaping in the air, face to face, flapping their wings so loudly that I could hear them from the veranda, several hundred feet away. They also thrust out their spurs just like the half-tame male pheasant that wandered in several autumns ago. What appeared to be fierce fighting continued for about a minute, the toms gobbling all the while, when suddenly they stopped and walked sedately, one behind the other, slowly up the field, stopping frequently to peck for food.

I sat watching from the stone wall beside the shed for over half an hour as they fed amiably together at the edge of the woods. Both had short, stubby beards and as they fed they constantly looked up vigilantly. Finally they ambled out of sight, still friends.

MAY 6. Early in the morning a rabbit hopped toward me on the driveway, but it was so intent on seeking food that it didn't see me until the last moment. Was it a female who had just had a litter of young? I wondered. The female usually leaves her nest just after the birth of her young to eat and mate again.

Then later, in a rainstorm, a female cottontail rabbit came tearing into the yard closely followed by two males. The first mounted her quickly from behind and then kicked her away with one of his hind legs. She raced off around a circular fence that encloses an enormous lilac bush and reappeared twice, each time hotly pursued by the second rabbit. All the while the rain fell steadily.

MAY 7. The Far Field area now has two allurements—the possibility of seeing a white woodchuck, red foxes, or both—so I walk there every day no matter what the weather. Today I was finally rewarded: the foxes were running all over the hillside above the old den.

As I sat in the black locust grove, the two adults and three kits alternately played and rested. Then, because it would only be a few days before the emerging leaves would totally obscure my view, I decided to move slowly out into the open. The only reaction was a quick glance from one of the adults who then trotted up to relieve the other adult who had settled down to sleep. That one jumped to its feet and loped off up the hill while the first one took its place, promptly lay down, and also closed its eyes. Two of the kits went below, but the third sat with its back to me before it too returned to the den, leaving the sleeping adult alone outside.

Had the foxes really seen me? Had they, in fact, been aware of me for weeks? They had often looked intently in my direction and then gone about their business. David Henry in Canada discovered that his foxes would allow a certain critical distance between him and them without fleeing, and I wondered if mine had established a similar, albeit much farther, critical distance in their relationship with me.

I was so pleased to find the foxes back in their old den that I returned for an evening visit. One adult lay outside the main den entrance, its tail tucked around its body, appearing to sleep but looking around occasionally. As I sat watching near dusk, the field was suffused with wood thrush music that washed over the peaceful evening scene like a benediction.

MAY 8. Walking along the Laurel Ridge Trail this splendid morning I detected the faint whiff of an alien odor drifting up from the farm valley below. It smelled suspiciously like the herbicide routinely sprayed on our power line right-of-way for several years until I persuaded the power company to cut the brush rather than spray it.

Today the odor permeated the air and caught in my throat. Why must the clear May air be saturated with poison? Because, I answer myself, using the land in whatever way necessary to make a living is the unspoken creed of most human beings. The facts that there are more of us all the time and that we are busily destroying the planet we depend on have not really sunk in despite the obvious misuses of the land that can be observed everywhere in the world. Although the number of enlightened people is growing, the majority still prefer to get as much as they can for themselves—and damn posterity.

But there are other, gentler ways to live, and some of those ways, like diversified and organic farming, are actually providing sound economic benefits to the practitioners. They are nonconsumptive, however, and our state, like many states, is more concerned with jobs in the here and now—such as mining coal, clear-cutting timber, producing still more glass and aluminum containers, manufacturing pesticides and herbicides, building larger farm machinery—than it is in teaching all of us to scale back our wants before we overwhelm the air we breathe, the soil we farm, the water we drink, with more pollutants than our bodies can withstand.

When the state Department of Environmental Resources first started spraying for gypsy moths, they claimed that the sprays were harmless. Yet dozens of people called me about the dying hummingbirds at their feeders. We had no dying hummingbirds, but then we had not sprayed. However, I couldn't lay all the blame on the DER. The state was besieged by ignorant people who stormed the legislative doors with naive demands for action. If we can send a man to the moon we can beat the gypsy moth, they argued.

Funny that no matter how many times over the centuries nature makes it clear that it cannot be beaten, humanity continues to battle on, usually to its own, as much as to nature's, detriment. "When will they ever learn?" as the protest song goes. But time is running out. We now know the consequences of our actions. Before we merely created a desert here, a polluted river there. Now we can destroy life on earth with a couple of buttons.

What we need are vast numbers of people whom the theologian Harry Emerson Fosdick once called "throwaheads," visionaries far ahead of their times. They must be people who have progressed from thinking about what the land will do for them to what they will do for the land, or from users to appreciators. Some of us fall in between, trying to live in harmony with the land, using it without harming the environment.

Every day we receive pleas in the mail for money to save the rain forests, save the whales, save the Arctic National Wildlife Refuge, save the spotted owl, save the old growth forests, save us from poisoning ourselves—the litany is unbearably poignant, the needs undeniably great. But a small minority of us are trying to pay for the stupidity and greed of the majority. How do we do it? How do we fight such venality?

Not everyone is susceptible, as I am, to the beauties of the earth, so they see no reason to save what they do not appreciate. Even as the poison odor drifted up from the valley, I could not help but rejoice in the delicate leaves once again emerging in all their miniature loveliness. It was a day to notice their beauty and color. The red oak leaves were deep red, soft and heavy-looking, reminding me of the fine old velvet used to adorn Victorian settees. Chestnut oak leaves had the look of ancient, beaten copper, while striped maple leaves appeared from their elongated, velvety, rose-colored buds a shiny, brassy green. The Hercules' club *(Aralia spinosa)* leaves erupted from a huge terminal bud at the tip of each large, upright branch, looking like large funeral urns top-heavy with leaves. Oh, it was a bonny day in May, and I pushed from my mind visions of a wasted earth, no longer capable of supporting the masses of humanity that ultimately depend on its life-support system.

MAY 9. I spent an hour in the mertensia patch, watching the creatures attracted to the long, blue-and-pink, tubular flowers swaying gently in the breeze. After only a couple moments of waiting, a ruby-throated hummingbird came whirling up to probe its beak deeply into each drooping blossom. It flew so close to me that I didn't need binoculars to tell that it was the ruby-throated male and not the pale-throated female.

Mertensia virginica or Virginia bluebell is a wildflower usually found in wet lowlands. But when my father presented me with several from his backyard, I planted them in my shady garden amidst the smooth yellow violets, wild ferns, and domesticated iris, and they spread like native mountain flowers. A member of the Borage family, our mertensia is one of forty species of mertensia worldwide.

Because it blooms early it is a popular source of nectar and pollen for our plush yellow-and-black-bodied queen bumblebees. They gather the pollen and nectar that they mix into bean-sized loaves on which they will lay a few tiny eggs. Those eggs will then be covered with wax. Queen bumblebees will also make wax honey pots the size of small thimbles which they will fill with honey to feed on while they brood their eggs. In this way the eggs remain warm until they hatch into bee grubs, which immediately burrow into the loaves of "beebread" and feed themselves. Eventually the grubs spin cocoons, change to pupae, and finally emerge as daughter bumblebees. They do all the further work of gathering nectar and pollen and making more beebread while their mother continues to lay eggs.

Yellow jackets of the family Vespidae, and cuckoo and carpenter bees, both solitary species of the social bee family Apidae, also frequented the mertensia patch. In addition there were various species of ichneumons or parasitic wasps, one of the largest families in the Insecta class. As larvae they often live on noxious insects and are valuable to have around. So are the syrphid or flower flies, which resemble either wasps or bees in appearance. They even buzz, but they do not sting. In larval form some of them feed on aphids, but when they become adults they prefer flower nectar, making them valuable pollinators.

Despite the large number of stinging creatures in the mertensia patch, I was not bothered as I sat there. Instead, I was surrounded by loud, industrious humming, part of the music of May which I absorbed as I watched the hummingbirds feast.

MAY 10. Once again our yard vibrates with the rollicking, unceasing, loud songs of a male northern (Baltimore) oriole. David asks irreverently, "What big mouth is making all that noise? I can't hear my music above it." For a bird to outdo David's blaring blues records is some feat, but the northern oriole is up to it. Throughout May the striking orange and black male pipes his whistled notes without end until he attracts the yellow and greenish brown female. The female weaves her pendulous nest while the male continues a muted singing, sometimes joined by the female. Then the noisy become silent and I rarely catch even a glimpse of them throughout the summer months, so much so that I often think of them only as May birds. Yet, after the leaves drop in the fall, I always find their swinging cradle suspended from the highest branch of a black walnut or black locust tree.

The eastern wood pewees also returned today and reminded me of the hot, summer days ahead when they drawled their slower rendition of the phoebe's song—"pee-a-wee"—which they sing throughout their time with us. Eastern wood pewees are definitely look-alikes of their close relations except that they have faint white wing bars and eastern phoebes do not; conversely, phoebes dip their tails slowly up and down when perched but pewees do not. They also differ in their habitat preferences. Phoebes nest in dooryards, outbuildings, or under bridges, and pewees live exclusively in the woods, their open nests built atop horizontal branches. And while the phoebes are among the first spring arrivals, eastern wood pewees are among the last.

MAY 11. I spent a good deal of time trying to track down an elusive warbler call to the left of Laurel Ridge Trail. But at last I found the singer. It was a hooded warbler, a species I usually find there at least once each spring, but unfortunately I can never remember its brilliant "weeta wee-tee-o" song from year to year. Male hooded warblers are satisfying to identify because they are so definitely hooded warblers and nothing else. Their distinctive black hoods that completely encircle their yellow faces and foreheads like a monk's cowl, offset by their yellow bellies, are unforgettable field marks.

As I poked about in that almost parklike area, crisscrossed with a maze of deer trails leading off seductively in all directions, I scared a hen turkey off her nest of twelve large white eggs on the ground amidst the laurel. Far from any source of water, the habitat that hen had chosen did not meet the specifications wildlife managers insist turkeys need.

A froggy-buzzy song along the Far Field Road alerted me to the second black-throated green warbler of the spring, followed by a close-up view of a magnolia warbler with his yellow breast, black streaks, gray head, white and then black around his eyes, and white wing bars. The woods were also filled with the singing of blackpoll warblers, and I stopped to watch two male cerulean warblers chasing each other, followed by two pileated woodpeckers doing the same thing. But for sheer visual beauty nothing could compete with the sighting I had of two male indigo buntings and a male scarlet tanager sitting together on a fallen log above the thicket. I felt as if I had been transported to the tropics where brilliantly colored birds are the norm rather than the exception.

Down in the thicket I finally settled myself for another spring ritual—watching the birds that come into the low growth at close quarters. The late naturalist, Sigurd Olson, in his book *Listening Point,* wrote that "everyone has a listening point somewhere . . . some place of quiet where the universe can be contemplated with awe." My own special listening point is the thicket where I can tuck myself in some small corner and watch the wild world around me. During a lovely May morning, it seems the best of all possible places to be, and I sat in an aromatic bed of gill-over-the ground, my back against an old, shrunken, cherry tree.

Trees in the thicket are few. Most of them are small and dragged down by the weight of wild grapevines. Hawthorn and red elderberry compete with a tangle of blackberry canes for space, along with a small grove of black locusts, an occasional red maple, and one flowering dogwood. Gaunt, dark remnants of dead trees brood above while moss-covered stumps and long-fallen tree trunks slowly rot into the soil. A thick carpet of mayapple leaves shines in the sunlight; in a moist area where a stream begins, lush grass is studded with purple violets.

All about me, birds sang and called. I listened in wonder to the music of wood thrushes and cardinals, rose-breasted grosbeaks and scarlet tanagers. Not so musical were the harsh cries of great crested flycatchers, the "chewinks" of the towhees, the raucous calls of blue jays, and the wild yelps of red-bellied woodpeckers.

Warblers were the stars today, however, and by sitting still I enticed them to land within arm's reach. Some saw me and fled. Others watched me intently for a few seconds, decided that I was harmless, and continued about their business of singing or hunting for food right in front of my delighted eyes.

American redstarts, both male and female, were particularly bold and often flitted in close to court and sing. Every bush, it seemed, had at least one male serenading a female. Actually, serenade is a charitable word for what most warblers do in the way of singing. By no stretch of the imagination are their songs beautiful. Many are variations on a theme of buzzing, like, for instance, the "song" of the golden-winged warblers. A male, with his gray back, golden head and wing patches, and black throat was striking to look at, but his song, "bee-bz-bz-bz," was monotonous and unmelodic.

Then a black-throated blue warbler landed silently on a nearby bush, and I was able to study him closely as he searched for food and occasionally rendered his wheezy song. He also did battle with a long, fat, beige-colored worm, jerking it about in his bill or seeming to reconsider by repeatedly laying it down on a log, only to snatch it up, jerk it some more, and finally swallow it whole. A black-and-white warbler crept along low-hanging tree branches while the olive-colored, worm-eating warbler with its black-and-buff striped head foraged in the dry leaves.

Just when I thought I had seen all the warblers there were to see I spotted two more skulking in the underbrush. Both were species I had never seen before on the mountain. The first was the rather undistinguished male Tennessee warbler—white eyebrow stripe; gray head; white neck, breast, and belly; and greenish back. The other was the more striking Kentucky warbler, looking somewhat like a common yellowthroat except that it had black sideburns rather than a full mask and a yellow eyebrow line that encircled its eye. Both birds were silent as they foraged only a dozen feet from where I sat.

When at last they flew off, I rose on cramped legs and wandered home along Sapsucker Ridge where the woods were filled with red-eyed vireos singing in force. One was even close enough for me to see his red eye. A dull-colored member of the vireo family with olive gray back and white breast, it makes up for its undistinguished appearance by its unceasing song that continues hour after hour with long pauses between each phrase. BORE, which stands for "Boring Old Red Eye," is its current birders' name, but I prefer the more imaginative and older nickname "The Preacher," suggesting someone who never stops sermonizing. Probably the longwindedness of old-fashioned preachers is beyond the experience of modern, yuppy birders, hence the new name that doesn't impress me at all.

MAY 12. A gorgeous day which I had to spend at the university. But I had my reward at dusk when Mark suddenly shouted, "Mom, Dad, come quick. There's a bear in the herb garden."

We were already preparing to shower. Dressing quickly, we hurried downstairs. The herb garden is directly outside our bow window, so we had visions (true, as Mark later reported) of a bear peering into our sitting room. But by then the bear had ambled down the slope, and we watched as it moved slowly past the old apple tree, out across the flat area, up into the woods along the Short Circuit Trail, and finally straight over Laurel Ridge. It was a medium-sized bear and coal black, like all the bears we have seen here.

Both David and Mark took off barefooted in the direction it had gone, David hoping to catch a glimpse since he had been down in the guesthouse listening to music and had not heard Mark's frantic call. But although the bear seemed to be moving slowly, in reality it traveled much faster than our running boys and they never did catch up to it.

I had been wondering for weeks whether we had a bear in the vicinity. Early on the evening of the fire Bruce and I had taken a walk and had noticed a black animal at the top of the power line right-of-way on Sapsucker Ridge, but it was just enough over the crest so that even through binoculars we could not get a good look. We assumed then that it was a dog, especially after we learned that people had been up on the ridge at that time.

Several nights later we heard a noise on the back porch, but when we went to investigate we found only that an old pillow, which we had put out in the trash, had been torn to shreds and spread across the slope. Another night I was awakened by the sound of my canning pot rolling down the porch steps.

Still I didn't put two and two together until Bruce and I discovered a large pile of bear scat on the power line right-of-way a few hundred feet from our house on the Laurel Ridge side. And several mornings after that I noticed that large strips of wood had been ripped from the power pole at the top of Laurel Ridge. With the actual appearance of the bear all the clues finally fit neatly together. We hope that it will return.

There is no excitement comparable, in an eastern North American woods, to encountering a black bear. Somehow it almost compensates for the loss of such magnificent animals as the cougars and woodland bison that used to inhabit the wilderness before it was tamed by white humanity. To see a bear in the wild is to feel, for a time, the atavistic fear mixed with wonder that our ancestors felt. My mind always says, "Black bears do not attack human beings unless they are sows cornered with young," but my emotions always caution, "Be careful. With an enormous, powerful animal like that you never know." Needless to say, I do not talk quietly to bears when I see them but step respectfully aside.

MAY 13. At 8:30 this morning I was once again sitting out in the open at the Far Field watching the foxes. An adult came trotting up to the den with a small animal in its mouth. Two yawning kits emerged but instead of mobbing or begging, they both examined the ground rather than the dead animal, even after the parent dropped it on the ground. Only when the adult approached one kit did it immediately pick up the carcass, but it soon dropped it again. Neither the adult nor the kits paid any further attention to it, contrary to all known wisdom about the feeding of young foxes which claims that the alpha pup usually gets first chance at any food, followed by the other kits in order of dominance. Instead they glanced over at me a couple of times, and then the adult settled down while the kits explored, rested, and played.

When the parent sat up abruptly, the kits instantly paused to watch as it moved restlessly around. Once it lay down again, the kits relaxed. One apparently wandered too far from the den site because the adult suddenly jumped up and walked over to the kit. Instantly it scurried back to the den entrance. That, I assumed, was an example of parental discipline, probably enforced with low sounds that I could not hear.

Finally the adult stretched and rubbed its back on the ground, which seemed to be a signal for the kits to play with it. One worried the adult's tail while the others jumped on its body. Eventually the adult rose, looked around, and trotted off. The kits continued exploring, and the alpha kit practiced a spring just once before they all settled beneath a shady tree to sleep in the open above the den site.

At last I got up and moved slowly off. They never noticed or, if they did, they weren't concerned. I could almost imagine them thinking, "No problem. Just that strange human who insists on watching us."

MAY 14. Several years ago we gave David a microscope for Christmas. As frequently happens in our family, the microscope became a family possession, and first David and then Steven spent a couple of weeks using it, calling me to have a look at their various finds. But the novelty wore off, the microscope was relegated to a desk drawer, and both boys moved on to other interests.

Today Mark resurrected it and spent hours painstakingly putting drops of pond water on slides, fiddling with the focus, and immersing himself in the almost inconceivable world of the minute. I teased him about getting "microscope eye" from peering through the eyepiece for such long periods of time, but after a few looks and a listen to his explanations of what I was seeing, I became a convert to his obsession.

He sat me down at his desk and gave me a lesson on the microscopic inhabitants of the wood frog pond. Already it is covered with green and yellow algae and is so crowded with tadpoles that a single scoop will bring up hundreds. With my eyes I can see the tadpoles, water striders, beetles, and even the minuscule water fleas. I can sit and watch the newts prey on the tadpoles and the tadpoles nibble the algae. In my ignorance, I thought I was observing the whole life of the pond. How wrong I was!

First of all, we looked at the two kinds of algae our pond has produced. Algae, because they perform 90 percent of the photosynthesis of an aquatic society, are very important to pond life. They are also valuable as food and as oxygenators of the water. No matter what their color (and their names do not always indicate their color) they do have chlorophyll. So far this spring, Mark has discovered two kinds of green algae, spirogyra and zygnema, and six kinds of yellow algae in our pond. All algae are unicellular (one-celled) plant life, but often the cells live in clusters or strings.

The diatom family of yellow algae is particularly interesting because of its diverse forms. Today we saw, for instance, the fan-shaped meridion diatom; the liplike cymbella; the long, thin fragillaria that have a bulge in their middles; and the disc-shaped navicula. These microscopic plants floated about looking very much like animals.

But then a protozoan, the paramecium, appeared beneath

the lens. This pioneer of the animal world was fat and spun in a circle as it moved about. We also discovered what some biologists believe to be a representative of the merging of the plant and animal worlds: the vorticella. Its cilia, or minute hairs, which together with its body contractions and expansions enable it to move one-tenth of an inch a second, make it the fastest of all the one-celled animals. It eats bacteria and paramecium, among other things, and it looks like an inverted funnel with hair around its edges.

From one-celled plants and animals we progressed a step upward to multicellular organisms often called wheel animalcules or rotifers. They are the chief components of the spring plankton bloom over ponds. Usually they are colorless, which makes it easy to see through them and observe their nervous, excretory, and reproductive systems. Their crown end with its cilia looks almost like a head, and often at their other end is a tapering, footlike appendage with sometimes a residual toe or two. We saw two kinds of rotifers: the plump, elongated asplanchna and a dicranophorus, the so-called tiger of the rotifers, that hunts its food through tangles of water plants.

I could see, as I peered down into that strange new world, why Mark was so fascinated. And today I cheerfully contracted "microscope eye" myself, so much so that when I looked up into my usual world, I was disoriented for a moment, still lost in the mysterious world of the miniature.

MAY 15. It looks as if we have a fortune in mayapples here on the mountaintop. According to scientists, mayapples may be an important source of an anticancer drug.

The mayapple *(Podophyllum peltatum)* is also called "mandrake," "duck's foot," "ground lemon," "hog apple," "Indian apple," "raccoon berry," "wild lemon" and "umbrella-leaf," almost all of which refer either to its distinctive, shiny green, umbrellalike leaves or its large yellow fruit. That fruit ripens in late summer long after the single, waxy, white flower has bloomed in May and the leaves themselves have fallen to the ground.

The alternate name "mandrake," however, points to its medicinal properties. Like European mandrake *(Mandragora officinarum)* the mayapple has a root that was used by the Indians in moderation as a purge; if they took too much, sometimes intentionally as a suicide potion, they would die. According to the old herbalists, the roots were ready to harvest soon after the fruit had ripened. Its active principle—podophyllin—is similar to mercury in its effects on the human body. *The Rodale Herb Book* states that it can be taken in small, frequent doses to "stimulate the glands to healthy action." But John Lust, in *The Herb Book,* warns that "if used during pregnancy, mandrake may cause birth defects in the child." Definitely not an herb for amateurs to fool around with!

But Indians had other uses for the plant. They claimed that an extract of mayapple rhizomes repelled potato beetles, and recent tests by the U.S. Department of Agriculture confirm this belief. In addition, Penobscot Indians used it to cure warts, while the Cherokees killed parasitic worms and treated deafness with it. Shortly after 1820 claims were made about the mayapple's effectiveness as a treatment for venereal warts, tumors, and polyps. But only recently were those claims taken seriously.

In 1984 a pharmaceutical company developed a mayapple derivative that is being used successfully in treating testicular cancer after chemotherapy has failed. In fact, the mayapple may be a cure-all—researchers have recently discovered that it also seems to fight such viruses as herpes I, herpes II, influenza A, and measles.

So today, as I walked our woodland paths and admired the blanket of twelve- to eighteen-inch-high umbrellas spreading over the ground, I wondered what other uses this versatile plant has and how many other plant species, still undiscovered, will turn out to be just as important to humanity? Certainly economic botanists have a rich field to mine, not only in the tropical rain forests where entire plant species are being exterminated before scientists can even discover them, but in our own country where the ubiquitous, but beautiful mayapple still thrives in deciduous woods from Quebec to Florida and as far west as Minnesota, Kansas, and Texas. It continues to be a source of wonder to me that for every disease humanity contracts, there seems to be a cure somewhere among the plants that inhabit our earth—proof of the intricate interrelatedness of the world that we continue to disrupt at our peril.

MAY 16. All through the early spring I worried about our chipmunk population. Usually I see the first ones on a warm day in February, and by late March and early April they are "cuck, cucking" away in the woods. But this spring I saw none until the first of May, when I finally spotted one sitting in the sunshine of the Far Field Trail. Then this morning I saw another, several hundred feet further along the trail, just beyond the bend. Both looked newly minted in their bright, reddish brown coats, and I wondered if they were the first of this years' litter that were born in early April. In Pennsylvania young chipmunks do appear aboveground in early May and are weaned between five and seven weeks of age.

On our mountain the chipmunk population fluctuates wildly, and the last couple of years it has been low. Some autumns we have counted as many as fifty diving into our fifty-two road drains as we drove up the hollow; other autumns we have seen only a very few.

Eastern chipmunks, *Tamias striatus,* are one of the easiest of mammals to observe aboveground because they, like us, prefer to be abroad during the daylight hours. I have often sat in the woods and had them chase past me or even run over my legs in their haste. That chasing is not playing, as one would imagine, but the highly solitary and territorial chipmunk defending its turf from invaders. So is the continual "cucking" they do in late spring and early autumn, the two dispersal times for the two litters most chipmunk females have in our area. This general singing, called "epideictic vocalization" by scientists, is done by the whole population of established chipmunks and apparently informs the dispersing youngsters of the population density of the area and how far they must go to find a less-populated place to settle in.

One researcher, Lawrence Wishner, claims for his study area of thirty acres in suburban Maryland a density of between 10 and 17.3 chipmunks per acre. Other figures, for our area and habitat, are 25 to 30 per acre. That seems hard to believe, but because chipmunks are near the bottom of the food chain—the preferred food for hawks, weasels, foxes, and snakes as well as dogs and cats—high numbers assure adequate food for all those predators. Chipmunks, in turn, eat not only vegetable food like wild nuts, mushrooms, and berries but an occasional salamander, bird, mouse, and small snake in addition to invertebrate prey such as slugs, snails, earthworms, insect larvae, and butterflies. Their cosmopolitan tastes should keep chipmunk numbers steady because there is never a time when *all* those food sources fail. Nevertheless, the numbers fluctuate wildly here. Just another mystery that neither the biologists nor I have solved yet.

MAY 17. The male gray catbird returned two weeks ago and sang to defend his territory from all rivals. His singing is a less polished mimicry of other birds, similar to brown thrashers. It is easy to distinguish a catbird from a thrasher, though, because the catbird frequently intersperses his nonstop imitations with the distinctive "meow" of a cat—hence its name.

The female, which looks exactly like the male, returned a week later, and after chasing through the shrubbery, she was ready to mate and start a family. Two days ago I noticed what I presumed to be the female catbird, since she builds the permanent nest, carrying a twig in her beak toward the grape tangle, and I stood watching to see where she landed. She saw me and remained perched on a vine until I turned away. Quietly I peeked and noted where the vines rustled. Several hours later I went to investigate the spot and found a couple of sticks lying haphazardly in a tangle of black raspberry canes about two feet from the ground. Could that be the beginnings of a nest or was it merely a trial building such as both male and female catbirds often do? Near dusk I had my answer. The twigs were covered with a piece of clear plastic arranged into a nest shape. That clear plastic has been the identifying characteristic of other catbird nests I have found over the years.

Between rain squalls today I ran out to check the progress of the catbird nest. It seems to be complete now, and the male is sitting at the top of the apple tree singing. Somehow the female labored through the pelting rainstorms of the last few days and finished the job in record time. Most observers have reported that it usually takes five to seven days of work to build a catbird nest.

MAY 18. Our old-fashioned, heavily scented, purple lilac bushes have bloomed with unusual splendor this May. Watching the magnificent, tree-sized bush that envelops half our front porch has become a favorite pastime. Butterflies of several species frequently visit it—particularly the striking black-and-yellow tiger swallowtails—and dozens of bumblebees work the blossoms from early morning until dusk.

In addition, it has been serving as headquarters for our local ruby-throated hummingbird couple. At first, I saw only the male zooming from flower to flower, his ruby throat glowing in the sunlight. Then, early this evening, I was treated to a breathtaking display of his rapid, U-shaped, courtship flight. Like a high speed pendulum, he swung three times over the lilac bush before diving down out of sight.

A few seconds later, the female appeared from the depths of the bush where, in all likelihood, she had also been watching the male's display. She fed on the flowers until dark and was buzzed twice by another hummingbird that chittered and then moved off so quickly that I could not see if it was the male or another female.

During previous springs I had noticed that lilac blossoms were attractive to hummingbirds, but it was David who, several evenings ago, informed us of the sphinx moth invasion. We all rushed out to look. In the dusk, we could faintly see the shadowy forms and hear the whispery sounds coming from the hundreds of moth wings that whirled about the bush. Mark caught several in his hands so we were able to identify them.

That first evening, having come out rather late to the event, we discovered only the hog sphinx moths—two-inch-long, brownish gray creatures with orange brown hind wings. But high above our heads we could see larger sphinx moths, although it was too dark to make out their colors.

Forewarned, though, is forearmed, and the following day I spent a good deal of time looking for day-flying sphinx moths on the lilacs. There are 124 species of sphinx moths in North America, and many come in striking color combinations, unusual for the mostly dull-colored moth tribe. I was shortly rewarded for my efforts by the discovery of one of the more flamboyant of the sphinxes, the two-and-a-half-inch-long hummingbird moth that has two maroon bands around the lower end of its fat abdomen and large, whirling, reddish wings so much like real hummingbirds that Roger Tory Peterson, in the latest edition of his *Field Guide to the Birds,* pictures the moth next to the bird. Almost every year, in fact, I can depend on at least one phone call from someone asking about an incredible creature that looks like a hummingbird buzzing around their flowers.

Luckily my close scrutiny did not bother the moth, so I was

able to watch it closely as it continually plunged its curved proboscis (two and a half inches long) into each lilac blossom. Sphinx moths are known for their unusually long tongues, or proboscises, much longer than their bodies in some species.

By eight o'clock that evening I was seated on the front porch, joined shortly by Steve and David. Mark stood down below the bush, ready to snatch and bring to us any new species we spotted. The first to appear were the two-inch-long, striking nessus sphinx moths. Their abdomens, circled by two bands of bright yellow, end in characteristic fanlike tufts. As the dusk deepened, Abbot's sphinxes became common. They were almost three inches in length and brown-colored with a wide band of yellow on their lower wings. Finally, just before dark, the hog sphinxes again appeared in large numbers. Since then the same species have appeared in the same order each evening. Tonight I watched the sphinxes as well as the true hummingbirds feeding, feeling akin to the Persians, who, according to plant historians, liked to spread their rugs before a single lilac bush so that they could meditate, pray, play their lutes, and eat sherbet. Then they would depart, refreshed by the beauty, fragrance, and mysteries of nature.

MAY 19. Today was my Big Day, the day I had designated for counting bird species on the mountain. I had set no alarm, but I was awakened by the singing of a persistent common yellowthroat. Instantly I tuned my ears to birdsongs and rapidly ticked off rufous-sided towhee, ovenbird, and wood thrush—enough incentive for me to rise despite the lowering skies.

Stretched out on the floor doing my daily exercise routine, I picked out red-winged blackbird, eastern phoebe, and song sparrow. Then I stepped outside on the veranda to check the nesting house finches at the top of the corner column, the starlings in the black walnut tree, and the robin threesome that had just hatched in the lilac bush nest.

Dozens of barn swallows were already sweeping the sky for insects, and a male bluebird was busily gathering food for his offspring in my bluebird box near the driveway. Chipping sparrows buzzed along the edge of the field, great crested fly-catchers "wheeped" in the woods, and the resident red-bellied woodpecker rattled from the top of the old pear tree.

I prepared a breakfast tray for Bruce and me and carried it out to the front porch along with my binoculars. We ate in companionable silence as I scanned the black walnut and locust trees beside the porch and listened for new birdsongs. Brown-headed cowbirds were courting, and the two American crows that have adopted our lawn as foraging ground swooped in overhead, veering off when they noticed us. The nesting white-breasted nuthatches gave their usual nasal calls, while the eastern wood pewees could be heard singing in the woods. Blue jays screamed past along Sapsucker Ridge; a downy woodpecker quietly tapped on a locust tree. Then I picked out the song of an American redstart just before I spotted it flitting, butterflylike, in the red maples growing along the driveway.

Suddenly the day took a more exciting turn. As we sat finishing our breakfast, a Cape May warbler landed on a branch right at eye level, quickly followed by two Blackburnians and half a dozen bay-breasteds. What was happening, I wondered? Usually most of the migrating wood warblers confined themselves to the topmost branches of trees along the stream or on the trails. But so far this spring even the topmost tree dwellers have been favoring small trees and shrubs. I have attributed this to the colder than usual spring weather, reasoning that what few insects there are are seeking shelter close to the ground.

I quickly finished off my breakfast to the trilling of field sparrows and the persistent northern oriole and, as I started up the Guest House Trail, I added the least flycatcher to my list when I heard its distinctive "che-bec" along the edge of the woods. The ruffed grouse I flushed beside the trail gave me my twenty-ninth bird by 7:45 A.M.

It was still heavily overcast, with a smell of rain in the air, so I had pulled my slicker on over my jacket. I walked silently along our moss-covered trails, and for the first half mile I heard no new birds. Perhaps the warblers in the yard had been a false alarm.

Then I reached the power line right-of-way at the top of Laurel Ridge. As I stood watching, the scrub oaks shimmered with foraging black-throated green, magnolia, and yellow-rumped warblers. In the underbrush, white-throated sparrows vigorously scratched up the wood duff; American goldfinches tinkled past overhead; and far below I heard the song of a northern cardinal. There were thirty-five species on my list as the factory whistle sounded 8:00 A.M. in the town below.

I never hurry during Big Days. I prefer to watch as well as tick off species, and so I moved in a meander that allowed me to see the wildflowers at my feet, the white-tailed deer melting through the woods, and even a box turtle lumbering across the trail. Then a flock of black-capped chickadees bounced along, a blue-gray gnatcatcher snapped insects from the air, and one black-and-white warbler ascended a tree trunk in the same spiraling way that a brown creeper does.

Along the road to the Far Field I listened for the rose-breasted grosbeak that had been singing in the same area for several days; I was not disappointed. Indigo buntings were easy to spot in the locust grove at the Far Field, while several turkey vultures drifted silently overhead. Coming back along the same trail, I heard the wheezy robin song of the scarlet tanager and the "whoosh" of mourning dove wings, giving me forty-three species.

But, as I neared the First Field Trail, I was suddenly aware of an increase in bird chirpings. Moving along a narrow trail edged with black birch saplings, I was surrounded by warblers perched on branches at eye level and less than six feet away. They fed as if I were not there, and sometimes they paused to watch me even as I watched them. Bay-breasteds, both male and female, black-throated greens, chestnut-sideds, a black-throated blue, and two Wilson's warblers I identified, but mostly I reveled in the sensation of being near so many warblers. Farther down the trail I also heard the unmistakable buzz of a worm-eating warbler in the same place where I had both heard and seen one several days before. They too breed on the mountain, and last summer I had come close enough to a nest (although I never could find it) to induce the anxious female to perform her crippled-bird act.

A solitary vireo in the woods and a gray catbird in the yard put me one species short of fifty when I returned at 10:15 to start lunch. As I chopped the soup vegetables, I regaled the family with my bird sightings. I was so excited that Mark decided to join me for a late morning stream walk.

He waited for me down on the guesthouse porch that overlooks a small lawn, the stream, a butternut tree, and a huge white lilac bush. When I joined him he was already agog with warbler sightings. "Mom, they're everywhere," he shouted, pointing to dozens of warblers in the bushes and trees and waving his hand at the large numbers flowing up along the stream bed. Nine out of ten were bay-breasteds, and I wondered what had caused such an explosion of them this day. In fact, the glut of bay-breasteds became a joke. With their dark heads, beige neck patches, and chestnut throats and sides, they are one of our handsomest warblers. They are also long-distance migrants that winter in Columbia, Venezuela, and Panama and breed in Canada and the northeastern United States. They, along with the chestnut-sideds and black-throated greens, were the dominant warbler species on this Big Day.

Slowly we moved down the road looking for new warbler species. And we got them—a Canada with its collar of black on a yellow breast; a northern parula sporting its distinctive greenish back patch and reddish neckband; a hooded; and three northern waterthrushes, in addition to red-eyed vireos and one veery walking along the road. We also watched a Cape May warbler bathing in the stream, its chestnut cheeks on a yellow head its special identification mark.

We returned triumphant with fifty-five species, and as I put the finishing touches on lunch, I scanned the mertensia bed outside my kitchen for the ruby-throated hummingbird that appeared as if on cue.

David joined us for the afternoon and helped us to identify the first yellow warbler of the season flashing along the branches of our ancient Seckel pear tree. An all-yellow bird—the male has rusty red breast streaks—his "sweet, sweet, sweet, I'm so sweet" song makes him one of the easiest of wood warblers to identify from a distance.

We walked back along Laurel Ridge Trail to its confluence with the First Field Trail where there is a young growth of black birch saplings. The place was alive with low-foraging warblers. Mark went crashing through the underbrush after a female black-throated blue, while David stood transported by the magnolia warblers that were near enough to touch. Large numbers of bay-breasteds, black-throated greens, and chestnut-sideds also came close.

It was Mark, though, who spotted the red-breasted nut-hatch on Sapsucker Ridge. And I picked up a winter wren in the old dump area where it likes to forage. A red-tailed hawk harassed by a crow and one cerulean warbler increased my count to sixty-two by dinner time.

After dinner I took my last walk of the day, this time with Bruce. Up until then he had kept a low profile, fearful that I would drag him into the fray. But now he set off briskly for the fork in the road where a broad-winged hawk had been hanging out.

"First we'll bag the broad-winged, then a wild turkey," he

confidently declared. Reluctantly I dragged my tired body after him. There was no hawk to be seen, even when Bruce located its nest in the woods, so I wearily ascended Laurel Ridge for the third time, this time in search of the turkey. On the way up I heard the soft "tap-tapping" of a woodpecker and picked out a hairy in the dimming light. I had been upset at the lack of woodpecker species because I had identified only the red-bellied and downy by evening, yet I know we have pileateds, flickers, and hairies in abundance. They must all be nesting, since they had been silent the entire day.

Evening, though, brought them out. Long after the light had become too dim to see anything, I heard both a flicker and a pileated. My total finally was sixty-five species, but I know I missed several common birds. I did not find wild turkeys or black-billed and yellow-billed cuckoos, no brown creepers or great horned owls, no broad-winged hawks or brown thrashers, although I had identified all those species on the mountain over the last couple of weeks.

Had our birding son, Steve, been here, Mark reminded me, our Big Day count would have been far higher. He would have steamed up and down the mountain several times, roamed miles along the valley, and followed the river. Kingfishers, mallards, great blue herons, eastern meadowlarks, house sparrows, rock doves, killdeer, tree swallows, chimney swifts, and many other water and valley species would have been added to the list.

I was content to have kept to the mountaintop, to have watched more than counted, and to have seen the largest warbler migration in my life, leaving me with a remembrance of warblers to cherish forever.

MAY 20. Last May I discovered a hillside of jack-in-the-pulpits. I had ranged far over the mountain in quest of birds and instead had found a hundred "jacks," preaching to a congregation of thicket dwellers.

The plants on this hillside came in a variety of sizes, shapes, and color schemes. Some had two sets of leaves, others had one. Many had only one leaf with no jack-in-the-pulpit or inflorescence at all. I peeked under dozens of the hooded flowers and discovered that both the insides of the hoods and the jacks themselves varied in color. The spathe, or hood, was liable to be striped with maroon and green, green and white, or just plain green. The spadix, or jack, that stood straight up from the cluster of inconspicuous flowers at its base was either a shiny maroon or green. Many of the plants were quite small, while a few reached to my knees. They grew in clumps in some places, independently in others. In fact, variation seemed to be the one thing those wildflowers had in common.

Jack-in-the-pulpits, *Arisaema triphyllum,* are perennial herbs and members of the Arum family, the same as the skunk cabbage with its hooded flower. The moist, deciduous understory of eastern forests is their preferred habitat, and they can be found in suitable woods from Maine to Florida. Their leaves are divided into three parts very much like those of the trillium. From the base of the leaves pops the jack-in-the-pulpit flower, whether the plant is male or female.

And therein lies the real peculiarity of jack-in-the-pulpits. They can "choose" each year whether they want to be male, female, or neuter the following spring. Being only flowers, the "choosing" is based on the amount of sugars from photosynthesis their underground corms have managed to store up during the growing season from May to September. If they have had a very good year (i.e., stored up lots of sugar), they will be females because it takes the most energy to produce that sex. If they have stored up a modest amount of sugar, they will be males. If it has been a poor season, they will be neuter and send up only a small leaf. This ability to change sex from year to year is called sequential hermaphroditism and is a rare breeding system in plants.

One botanist, Paulette Bierzychudek, studied jack-in-the-pulpits for several years. Each September she dissected the underground corms of the plants to find out whether they would be male, female, or neuter the next spring. She also claimed that anyone with a little experience can identify the present sex of a jack-in-the-pulpit by the overall size of the plants. The largest plants, usually with two leaves, are females; the medium-size ones with one leaf (by far the most common) are males, and, of course, those single leaves without flowers are neuters.

This tentative identification can be verified by peering down at the base of the long, thin spathe. Clustered there are either the white or purplish anthers of the males or the green spherical female structures. In addition, if it is female, the bottom edges of the hooded spathe will overlap neatly. The male spathes have a small gap near the base. That gap allows the tiny fungus gnats, which cross-fertilize jack-in-the-pulpits, to escape from the male flowers after they are covered with pollen. But when they visit a female flower with their pollen, they are trapped there and die after doing their work.

The pollinated flowers produce fruits which enlarge and gradually burst out of the spathe. By summer's end the fruits consist of a cluster of scarlet berries, each of which contains one or several seeds. These fruits are eaten, to some extent, by pheasants, wild turkeys, and wood thrushes. However, the leaves and corms of jack-in-the-pulpit contain calcium oxalate which deters most herbivores from eating them, although I do find an occasional leaf or jack chewed down.

Today when I went to find that same patch only a few scraggly plants remained. That wonderful thicket had been carelessly logged last fall and winter, and few of the plants survived. This is the second time I have found a substantial patch of jacks on the mountain only to have the place stripped for lumber. That was why I was so glad, earlier in May, to find that one plant along our stream. Maybe it will multiply and give me a patch of jack-in-the-pulpits on our own land that will not be destroyed by a logging operation, at least while we have anything to say about it. There are some things more valuable than the money we could get from logging our five hundred acres. One of those things is a hundred jack-in-the-pulpits.

MAY 21. High in the treetops, the gypsy moth caterpillars have reached the one-inch delectable size beloved of cuckoos and cedar waxwings. Of course, I cannot see them although the steady rain of frass (caterpillar excrement) warns me that there are many and that they are growing fast. No, it is the sudden influx of both species of cuckoos and of cedar waxwings that tell me of the moths' progress.

Cuckoos always seem out of place here. Surely such birds belong in the south and west with their close relatives the mangrove cuckoo of south Florida; the groove-billed and smooth-billed anis of Florida, central Louisiana, and points as far south as Argentina; and the comical greater roadrunner of our Southwest. But no. Our yellow-billed cuckoo, which has the same white-spotted undertail as the mangrove cuckoo, nests here. So does the black-billed cuckoo with its smaller tail spots. The colors of their bills do distinguish them. So do their songs. The black-billed sounds more like a cuckoo should, with its fast "cucucu, cucucu, cucucu," sometimes sung even at night. The yellow-billed is deeper and longer, a "ka-ka-ka-ka-ka-ka-ka-ka-ka-ka-ka-ka-ka-kow-kow-kowlp-kowlp-kowlp-kowlp-kowlp" which drags out near the end.

One spring I found the carelessly built nest of a yellow-billed cuckoo cradled in the lowest limbs of a laurel bush. Since the shrub was next to one of our trails, I was able to watch it closely and observe the nestlings that reminded me of small porcupines because of their long, pointed feather sheaths which only burst into feathers after the young are half grown. That mother hatched two cuckoos but whether or not they reached maturity I never learned because eight days after they hatched they were gone. However, they were quite agile and usually feather out in a few hours during their seventh day, so they could have already been launched into the world when I paid one of my infrequent visits.

I have never found another nest despite the upsurge of cuckoo numbers following the arrival of the gypsy moths, but I have had plenty of close looks at both species. They hang around until fall when a native scourge, not nearly so destructive as gypsy moths, but more unsightly than tent caterpillars—the fall webworms—provides delectable eating. These cuckoo birds obviously prefer the hairiest and most repulsive caterpillars.

MAY 22. I have continued my watch for the foxes, and was finally rewarded today at noon. As I stood across the field looking at the hill den, an adult fox approached with a chipmunk in its mouth, hesitated in front of the entrance, and then, as if recollecting that the kits were no longer there, trotted on down the path toward the thicket den, which has been, over the last couple of weeks, collecting more and more reeking carcasses.

Twice it stopped to investigate something on the ground. Once it looked straight at me, exposed as I was, standing in the open and peering through my binoculars. But I never moved and on it went, disappearing behind the pile of brush that leads directly (a yard or so) to the thicket den as I followed it quietly on the well-worn fox path liberally strewn with fox scat. I heard nothing at the site, but the proof that they are using both dens seems irrefutable. It would make sense for them to desert the hill den because it is now so overgrown with weeds and grasses that reclining foxes would have no view of their surroundings. Also, it is extremely hot at this time of year and the thicket den is not only cooler but it is closer to the source of a small stream.

MAY 23. Unlike people in towns and cities, we do not consider the European starling a pest species. Usually we have only one breeding pair on the mountain, although last year there were none. So we were pleased when a pair took up housekeeping in a black walnut tree cavity high above the driveway.

I had a good view of the nest hole from my kitchen window and often glanced out at it, especially as the noise of the growing youngsters increased. Their parents worked at a terrific rate to feed them, and I was certain that the nestlings were close to fledging.

Then a couple of days ago, as I sat at the kitchen table eating lunch, I noticed that the starling parents kept landing at the hole, but instead of going in would spring back off the tree trunk and hover in the air.

"Something's wrong at the starling nest," I told David and ran to get my binoculars. By the time I returned, the curving branch near the hole had been occupied by the male parents of several neighborhood bird families. Lined up in a silent row were a red-winged blackbird, a house finch, and a robin. They bore an uncanny resemblance to human onlookers at a tragic accident.

The silence was eerie. Even the starling parents, who persisted in landing and then springing back off the tree, were quiet. I kept my binoculars trained on the hole and after several minutes, a snake's head appeared at the opening. At that sight the male starling uttered a low, guttural noise, a ghastly sound that, to my anthropomorphic ears, seemed to sum up the hopeless tragedy that had befallen their little family. Nevertheless, the parents continued to fly in with food all that afternoon. They would land at the hole, jump off, fly to a nearby branch, walk around for several seconds, and then eat the insects they were still conditioned to bring for their youngsters.

The neighbors returned to their own families except for the father robin. He kept a vigil there for the rest of that day and for several days thereafter. Whenever the snake tried to leave the hole, he would fly at it and the snake would quickly withdraw. Meanwhile his three nestlings, which are close to fledging, are guarded by their mother in their lilac bush nest.

I have kept a close watch, hoping to see the snake emerge, since I want to measure it and make certain of its identification. Nearly twenty-four hours after the death of the starlings it finally crawled out of the hole, but instead of coming down the tree trunk, it climbed out onto a long curved branch to bask in the sun. It was over five feet long and was a black rat snake, *Elaphe obsoleta obsoleta,* which is known for its ability to climb trees.

In less than five minutes father robin came flying toward it, diving repeatedly and giving its "tut-tut-tut" warning call. That snake was back in the hole in a couple of seconds. It may have had a good meal of starling nestlings, but the male robin was making certain it would not get anything else to eat for awhile, nor would it be allowed to relax in the sun. This cat-and-mouse game has continued for three days until today when the young robins fledged. Then the father robin withdrew.

The hole is now the summer home of the black snake. It probably lived there last summer as well, since black rat snakes like tree cavities. While I could empathize with the starling family, I could not condemn the snake for eating the young birds. Snakes, after all, are also part of nature's scheme, and on a farm they are extremely valuable in controlling the rodent population.

So I continue my role as curious observer, trying to make sense of what I see, but not passing judgment on the way that natural forces are balanced. Although I may not be as fond of the snake family as I am of birds, for me to interfere on the side of the birds would have been an arrogant decision that presumes my wisdom is greater than nature's. That attitude has been a part of humanity's thinking for far too long and has led to too many abuses. I am content to believe that the web of nature is too intricate for me to meddle with, even on so small a scale as destroying a black rat snake to protect young birds.

MAY 24. I have found only one member of the orchid family on our dry mountaintop, and that is the moccasin flower or pink lady's-slipper. Every year I count the ones I find growing beside our trail, and today I discovered a total of twenty-two flowers growing along the Laurel Ridge, Far Field, and First Field trails. Finding lady's-slippers from year to year is always a challenge because they have migrated from place to place during the years we have been here.

Our first year I found twenty growing clustered in a flat patch of woods a hundred feet from the guesthouse. The following year there were only a few blossoms there and a few others scattered up along the Guesthouse Trail. The year after that there were none left in that area at all. Instead, the boys discovered several along a path farther down the mountain which they renamed the Lady's-slipper Trail, but when I went to look the following year I did not find any.

That year they began to scatter themselves along what we were then calling the Old Dump Trail. Not liking that name, Bruce suggested that we rename *that* trail the Lady's-slipper Trail. Two springs later the lady's-slippers were gone from there, so the trail reverted to its previous name.

Recently the lady's-slippers have migrated to the very top of the mountain, growing only along the topmost reaches of the Laurel Ridge, First Field and Far Field trails. Why they should continue to move so radically (between a mile and a mile and a half over the years) continues to be a mystery to me, but as long as they blossom somewhere on the mountain every spring I am satisfied. Because they are liable to pop up almost anywhere, they seem to be more closely allied with birds and mammals than with stay-at-home flowers.

MAY 25. The youngsters are hatching. First I surprised a hen turkey in the high grasses of the Far Field with at least one poult. I had a close-up view of its heavy brown mottling on a light yellow background before it scurried off toward its clucking mother.

Then, as I neared the old dump walking down the First Field Trail, a mother ruffed grouse suddenly erupted at my feet on the path, leaving nine chicks huddled in a circle where she had been brooding them. All but one scattered instantly. I knelt to look closer at the remaining downy yellow chick with brown above its eyes that sat a few seconds longer before scuttling away to join its newly hatched siblings.

Because it was cool (fifty degrees with a strong breeze), I quickly left the area so the hen could reassemble her chicks and brood them. Judging from the chick I saw eyeball to eyeball, so to speak, they were little more than a day old. Although they were already mobile, they still needed protection from the unseasonable weather.

MAY 26. This glorious morning I drove down the mountain early to do some shopping. Halfway down the hollow I subconsciously swerved to avoid a pile of leaves on the road. Yet something registered in my mind, and I thought to myself, "That leaf pile was really a fawn." But I was in too much of a hurry to stop and check.

Two hours later, coming back up the road, I saw that the "leaf pile" had shifted to the left rut and that it was, indeed, a fawn. Closer investigation proved that there were two fawns, lying curled up and absolutely still. I got out of the car to persuade them to move, figuring that they would run as I approached. But they didn't. They remained motionless until I reached down to touch them. Then they leaped to their still-wobbly legs. One cried, "mah, mah," and I heard a rustling from the doe in the woods. But she did not appear to defend her twins.

Finally, as I stood waiting, one little fawn wobbled up the road and turned aside at the second pull-off. The other looked confused, stumbling and falling once before making it half-way down the slope to the stream. There it curled up again, attempting to blend with the dead leaves, but, of course, I knew where it was. Both were exquisite creatures, and I was pleased to have had such a close look at newborn twin fawns. I see them more frequently every spring as the white-tailed deer population increases due, in part, to the mild winters we have been having. No longer do deer struggle to survive through the snow and cold. Both food and shelter are plentiful, and winter has become, for them, almost an easy time; I often watch them playing tag down below our house even as a gentle snow falls.

MAY 27. Ever since I read Margaret Morse Nice's classic *Watcher at the Nest* I have dreamed of watching nests as she did. Yet, whenever I find nests in the woods—several years ago I discovered five wood thrush nests within a couple hundred feet of one another—and begin to keep an eye on them, one by one the nests are ransacked and within days all eggs and/or nestlings disappear. So I have concluded that nest-watching in the woods is simply too dangerous to the health and well-being of the little families, and when I do accidentally stumble on one I back away quickly, almost as if the nest contains poisonous snakes instead of little birds.

However, there is one bird whose nests are open to careful scrutiny without any danger to the nestlings, and that is the eastern phoebe. Here on our farm they always build their nests in secure places where predators cannot reach them. The guesthouse portico is particularly secure, and this spring, for the eleventh year in a row, the phoebes have rebuilt that nest.

Not only is their site impregnable, it is, unlike the other sites around our yard, easy for me to observe. I can sit inside the front door of the guesthouse and watch the family going about its business without disturbing them in the least. Phoebes are trusting birds. Even when I climb up on a chair and hold a mirror above the nest to check the eggs, both parents wait quietly and patiently on a nearby fence post until I leave. I did that once a day beginning on May 1 when I found five white eggs in the nest.

They began hatching on May 9. By 7:30 P.M. the nest held three eggs and two nestlings. The next morning at 7:15 there were four nestlings and one egg. Eventually that last egg was removed, probably by one of the two feeding parents, and the nest population remained at four.

Since the very young ones were quite helpless and almost devoid of feathers, they had to be brooded as well as fed for several days. During the first eleven days, I confined my watching to a daily mirror check, but by May 22 the nestlings were fully feathered, plump, and vigorous. That was when I began observing nest activity more closely.

I spent one hour each day on May 22, 23, and 24 recording the number of feedings. I found that, on the first two days, they were fed twenty-five times. I assumed both parents participated, but since adult phoebes look alike and since, during that time, they never appeared at the nest together, I couldn't be certain.

In addition to feeding, the parents also kept the nest and its surrounding area clean by catching, in mid-air, the nestlings' fecal sacs. There was a definite routine to this. A nestling would put its tail over the edge of the nest and wiggle it a few times. This alerted the parent in time to catch the sac in its beak and carry it off.

By May 25, when the nestlings were sixteen days old, the feedings had diminished to fifteen an hour, and the young birds spent a lot of time flexing their wings or standing on the edge of the nest and beating them. According to Arthur Cleveland Bent's records in his monumental, multivolumed *Life Histories of North American Birds,* phoebe nestlings should have been fledging between their sixteenth and seventeenth days, but the weather was wet and the birds seemed very timid.

When I went down yesterday to watch, I found the activity so engrossing that I could scarcely tear myself away. One parent appeared with nesting materials in its mouth and, despite screams of hunger from the chicks, proceeded to build a second nest right next to the first, even jerking nesting materials from the side and top of the first nest as the youngsters looked on. While the one parent worked on construction, another landed and fed the nestlings. Since the female always builds the nests, I assumed that she was the nest-builder and that the male was the feeder.

The mother clearly had two separate instincts battling within her because sometimes, when her nestlings screamed, she tried to shove the construction materials down their throats—which they promptly spat up. Also, although they made the proper defecation signals, she ignored them, so the pavement below the nest became messy for the first time. The male did catch them when he could but, by and large, nest and nestling care had become haphazard at best.

Slowly, the young birds began to take the hint, and there was much wing-beating on the edge of the nest. Then, even while the female worked on the new nest, first one nestling, then a second, jumped out of the old nest and into the new. The female paused and watched them, but I never saw her encourage them in any way. After hopping about in the new nest for a few minutes, they hastily returned to the remains of their old nest. By nightfall they were all still there.

Today I was down at the guesthouse by 9:30 A.M. As I passed under the portico, there was a sudden "whoosh," and out flew all the phoebes—fledged at last.

MAY 28. At dawn it was already seventy degrees and clear, and for the first time since last October I sought the shade rather than the sun to sit in at the thicket. But first, along the Far Field Road, I stopped to listen to a hermit thrush singing somewhere off in the woods, a privilege I have had only once before here. Then I found an indigo bunting nest in the same slanting blackberry shrub along the road where indigo buntings had nested three years ago. The nest was attached to the stems with the cottony remains of fallen black willow catkins, and the parents scolded as I bent to admire their workmanship.

After seventeen years of searching, I at last stumbled upon the nest of a rufous-sided towhee as I walked slowly through the thicket in search of a cool spot to watch wildlife. The female alerted me when she flew up literally at my feet, and after only a few seconds I located a nest containing three, almost round, mottled brown eggs. Constructed of strips of bark and lined with dry grasses, it was built neatly into the exposed root of a small Hercules' club tree. I moved quickly on so the female could resume her brooding and sat down amidst a maze of grapevines, my back against a sheltering tree. Several times I heard the muffled drumming of a ruffed grouse, but only a female redstart came in close to forage among the branches of a small, wild black cherry tree. After a time I rose quietly and meandered on down to where the water seeped out into the bare beginnings of a stream.

As I entered the moist area, I heard the unmistakable cry of a female ruffed grouse. She came sailing toward me, her tail fanned out. Then she crossed the stream and scrunched her body low, crying all the time. In the meantime a male, ruff extended and tail fully fanned out, strutted off into the underbrush, obviously the same one I had heard drumming while sitting in the thicket. The female continued to stalk around and whine, and then she climbed up into a fallen tree branch where she kept up a steady, henlike clucking as I watched for over half an hour. I never saw or heard any chicks. Neither did I find a nest, yet she acted like a grouse defending one or the other. If so, why had the male been engaging in the courtship display? Had she just lost her brood to predators and still retained her protective instinct even while she was seeking out a courting male? Was she interested in having a second brood before the first was raised?

Usually ruffed grouse females have only one brood a year. Once again I witnessed nature in action, and once again what I witnessed is unique to the literature as far as I can determine. But then it wouldn't be an Appalachian spring if I didn't have at least a couple of new mysteries to add to my store of unexplainable and undocumented natural occurrences.

MAY 29. Nature rarely repeats itself but this evening it did. At just the same time last night I had gone for a quick walk along the Short Circuit Trail, expecting to do nothing more than stretch my legs. But above the old dump I heard the rustling of a mature deer in the woods above me, and a few seconds later I discovered a newly born fawn sprawled out at the edge of the Dump Trail. It looked so young that I wondered if the doe had run the moment she dropped it, having heard me approach.

The fawn already knew enough to lie absolutely motionless, but its legs were still bent at awkward angles, its tiny hoofs dug into the earth. Only its large, blinking brown eyes signaled to me that it was alive. I knelt down to speak quietly to it, trying to assure it of its beauty and my peaceful intentions, before hurrying on so that its mother, whom I could still hear rustling in the woods, might return to it.

This evening a fawn was lying in the middle of the same trail, a few hundred feet from where I saw it last night. This time it was properly curled up, but otherwise it looked the same as last night's fawn. And again I could hear a doe in the woods above us. There may be creatures more beautiful than young fawns in this world, but if so I have yet to see them.

MAY 30. I sat at the edge of First Field in a bed of grass and listened to the unceasing song of the red-eyed vireo. A small beige butterfly twirled around me and stopped to investigate the salty sweat in the open palm of my hand. It paused and slowly pumped its wings up and down. Suddenly the ordinary, dull-colored butterfly was transformed. As its wings opened they revealed a pattern of variegated beige and brown that looked like an especially exquisite piece of handmade lace.

How often when I stop to look at something in nature that appears to be quite ordinary does it become, on closer inspection, extraordinary—reason enough, I've decided, to pause and let nature come to me rather than continually pursue it. To sit and be brushed by a butterfly's wings is not an experience to disdain.

"Be still and know that I am God," the Old Testament admonishes. Yet in our present-day, hurry-up society few people heed those wise words, and peace of mind has become rare. If I had one wish for the world it would be that troubled people everywhere could experience the beauty of an Appalachian spring day by sitting in a green meadow, as I did, and take the time to watch an ordinary kitchen maid become a dazzling Cinderella, her ball gown of lace her crowning glory.

MAY 31. Along the Laurel Ridge Trail this morning a doe and her fawn loped ahead of me. I froze, thinking they had seen me, but after a short distance the fawn, who was in the lead, spun around and ran toward me as if pleased to be running more energetically than its sedate mother, who followed slowly behind.

They paused to browse on the trail, then turned and ran up the trail again. A second time they whirled around and headed back to where I remained standing motionless, the fawn still in the lead. They stopped sixty feet in front of me, and the doe suddenly hesitated, her ears pricked up, and looked intently in my direction. Slowly she stomped each front hoof down in turn, signaling that she knew danger lurked in the dim sight or smell of me that she sensed. Then she turned and fled into the woods, followed by the confused fawn who ran several more feet up the trail before plunging into the heavy undergrowth that had already enveloped both the body and the sound of its mother.

Seduced by the lovely day I took a second walk in the afternoon. As I descended the First Field Trail, I again froze in my tracks when I spotted a doe and her awkward, leggy fawn wandering up the trail toward me.

Neither saw me, even when they paused at a seep, the doe to drink the water, her fawn to drink from the doe's teats. As the doe looked back at her nursing offspring, I quietly and carefully sat down by the side of the trail. There I watched the scene of fawn and doe unfold for over half an hour.

The doe licked her fawn while it nursed, and then both meandered closer to where I sat. Sunlight filtered down through the newly emerged tree leaves, leaves as fresh and pristine as the spotted fawn. While following the seep, the mother also nibbled on the plants beside it; the fawn seemed to be imitating its mother in its feeding, although it clearly preferred milk to greenery.

Occasionally the doe looked in my direction, swinging her head from side to side as if trying to catch my scent, while the fawn followed obediently at her heels. From an original distance of 150 feet, they advanced as close as 80 feet before the doe finally turned off into the woods, her offspring trailing behind. I never moved until they were gone, not wishing to panic them into headlong flight and thereby ruining the peace and beauty of the scene. It was only when I stood up and stretched my cramped limbs that I realized I had been sitting in the mud.

Both the morning and afternoon scenes, from beginning to end, were played out in utter silence—animated, soundless action that properly concluded the best of all possible months.

Climax

JUNE 1. "And what is so rare as a day in June?" wrote the New England poet James Russell Lowell over one hundred years ago. "Then, if ever, come perfect days." Every year I repeat those words as day after June day the thrills and beauty increase to an almost unbearable climax of joy before the flux of spring days end and the less active days of summer settle over the mountain in a haze of heat and humidity that drugs wildlife and humans alike.

In June there are still young creatures everywhere—new fledglings making distress calls, the lost cries of "mah, mah" from fawns, the scurrying off of young ruffed grouse while their mother remains to lure away predators. "The high-tide of the year," Lowell called June, when "we hear life murmur or see it glisten."

Today I watched it glisten—the yellow hawkweed shining golden in First Field, the wild rose bushes in our yard and field resplendent with pale pink blossoms. But, in reality, this was the day of the turtles. In the wet area beside First Field Trail a red-eyed male box turtle in his bright yellow-marked shell was on top of the larger, but less flamboyantly colored female and hissed at me as I paused to watch. Apparently he did not appreciate an audience while he mated. Actually he had been interrupted just as he was in the process of hooking his hind toes in the female's rear plastral opening, after which she would have clamped it shut over his toes. Then he will slip backward until he is sitting upright on the ground, or even leaning over backward, before finally mating. I did not stay to watch, afraid that I might discourage the lovers, but I wished them well.

When Bruce came home from work he reported seeing a snapping turtle plodding up our road a mile from the bottom of the hollow. It was heading over to the narrow trickle that our stream has become. That was only the second snapping turtle we have ever seen on the mountain—or rather that Bruce has seen; he saw the first one while driving up the hollow, only that turtle had been about a quarter mile up the road. Our shallow stream would not seem to provide the proper habitat for snapping turtles who like deep, still water, well populated with young ducks and fish, neither of which our stream has.

However, June is the time when female snapping turtles travel about in search of the right place to lay their eggs. They will go a considerable distance from their living quarters to find the side of a bank where they can scoop out the earth with their hind legs, lay approximately twenty-eight spherical eggs, seven-eighths of an inch in diameter, and then have the earth collapse down over the eggs as they crawl away. The turtle Bruce saw was definitely headed toward an ideal bank site.

JUNE 2. Too long I have been addicted to trail walking. Bushwacking on the mountain has charms I never dreamed of. I picked a faint deer trail leading down Laurel Ridge and meandered my way past laurel and blueberry shrubs this beautiful morning, finding enough open area to see where I was going.

I was accompanied by the penetrating calls of several competing ovenbirds, the buzzing of worm-eating warblers, and the flutelike songs of the wood thrushes. Ahead of me three deer filed past, walking at a steady, unhurried pace. They neither saw nor heard me and made me feel more a part of the woods as I slipped quietly along than I do when I am mistress of the open trails, where deer often leap away in snorting fright.

Then, too, although I knew the general direction I was headed, I didn't know what I would see or where I would end up. Such uncertainties undoubtedly added to the joy of bushwacking.

As I descended, the underbrush grew thicker and the mountain steeper. Even the singing birds changed: I now heard the unending drone of red-eyed vireos; the slow, drawn-out calls of eastern wood pewees; the nasal "wheep, wheep" of great crested flycatchers; and the repetitive, ringing songs of northern orioles.

Eventually I heard the gentle flow of our stream below and discovered still another well-defined deer trail that led me directly down to the water. I forded the stream, balanced precariously on a loose rock, scrambled up the steep and muddy bank on the opposite side, and discovered that I was almost halfway down our hollow road.

Walking back up the road, I recalled a conversation I had overheard one day when I was at the university. Three women were discussing a fourth woman's trip to Bermuda. "She said it was very boring," one woman reported.

"Well, at least around here we have a better grade of boredom," another replied.

It was clear that none of those women were at home in the outdoors, and I mentally compared them with friends of mine who are. One thing we never complain about is boredom. The outdoors always offers us new and exciting experiences, which is what I found this morning.

JUNE 3. I went back down to the stream this morning to accompany David who is using William M. Beck Jr.'s Biotic Index to rate the pollution level of our stream. From time to time he checks the numbers and species of aquatic invertebrates there to see how many pollution-sensitive creatures he can find. Depending on what he discovers, he can match it to Beck's work and see just how pure (or impure) the hollow stream is.

The stream drains only our mountaintop, and during heavy rainstorms at this time of the year we can find the several springs in First Field from which the stream originates. Since we have a relatively dry mountain, the stream is never more than four feet wide and sometimes, in time of drought, considerably less. In many places the mountain slopes on either side of the stream rise so precipitously that the only way to explore it is to wade through it as I have occasionally done.

Today, clad in sneakers, David and I headed down the streambed. He carried a potato rake, a piece of fine screening stretched between two poles, jars with screw-on lids, and tweezers. While I stretched the screen horizontally across the stream, he raked the bottom with the potato rake. Then he gathered up the screening and, using the tweezers, carefully picked up the minute invertebrates trapped by the screening and put them into the jars. He moved down the streambed, performing the exercise twice more. Then he began turning over rocks to capture still more creatures.

To the untrained eye, many of the things he picked up did not look very impressive. Some did not even look alive. For instance, the cases made by caddisfly larvae to protect themselves provide excellent camouflage. They spin silken tubes and embed pieces of leaves, sticks, grains of sand, tiny stones, or minute shells into them. The chosen materials are cut to size by the larvae and laid end to end or in geometrically woven patterns which are often intricate and beautiful. The swifter the stream, the heavier the constructions built by the larvae. They are relatively safe inside their cases and even pupate there. Finally, they emerge from the stream as small, brownish, mothlike adults. Some of the cases we found today were turtle-shaped, some were clusters of stone and vegetable matter, and others were hollowed-out twigs.

He also discovered four types of stonefly nymphs. They, like caddisfly larvae, indicate a clean stream and are considered of enormous importance as fish food. Stonefly nymphs are sluggish creatures which rest on the stream bottom or cling to the undersides of rocks, eating algae and plant debris. Ranging from six to sixty millimeters in size, according to their species, they live as nymphs in the water from one to three years, crawling out of the stream in early spring or late fall to transform into shade-loving adult stoneflies.

Mayfly nymphs, another indicator of a pure stream, were abundant, and David located three species. One clung tenaciously to the surface of a stone, the second was free-ranging, and the third was a sprawler which sat, covered with silt, on the stream bottom. These nymphs, which eat diatoms and other minute plant life, spend one to four years in a stream, providing food for the larger dragonfly larvae, beetles, and fish. Eventually they crawl out of the water and molt into graceful, short-lived mayflies that dance over a stream or lake, mate, lay their eggs, and die.

David also discovered two larger inhabitants of the stream: a bug-eyed dragonfly nymph and a crayfish. The dragonfly nymph is able to tolerate some pollution in a stream and therefore is not quite as important to David's study as the caddisfly larvae, crayfish, and mayfly and stonefly nymphs. But once he tallied up the results of his search, he was able to give our stream a rating of twenty-three points. According to Beck's index, any stream with eleven or more points is a clean stream.

If we can only prevent lumbering by our neighbors on the extremely steep, erodible slope above the road, the stream will remain just as pure as it was today. But despite a few weak laws on the books that are rarely, if ever, enforced in our state, there is little regard for stream quality during most lumbering operations, especially if the stream is too small to support such fishable delicacies as trout and bass.

What good is our small stream, given the user mentality of most people? Contrary to what Ralph Waldo Emerson said in "The Rhodora," beauty is *not* its own excuse for being—as far as users are concerned. The fact that our stream ripples quietly over sun-dappled rocks, flowing past second growth hemlock and beech trees, and provides sustenance for innumerable plant and animal species hardly matters in today's economics, at least in our county where jobs are scarce and the people are as materialistic as anywhere else. But when the beautiful places are gone, what will feed the psyches of such people?

JUNE 4. I spent considerable time in the mid-afternoon trying to locate the nest of a pair of obviously agitated field sparrows on the power line right-of-way, but instead of field sparrow nestlings I got turkey poults. A hen turkey suddenly appeared on the trail behind me, and I counted little poults everywhere, peeping loudly while she clucked softly. There were at least nine, and probably more, although it was difficult to keep count because she kept moving in and out of the underbrush and the poults kept erupting out of the brush to search for her. She certainly did not react to a human's presence in the frantic way that ruffed grouse hens do. Instead, she walked with all the bearing of a queen, her head high, her pace measured, casting her eyes about in every direction as she continued up the trail away from me. Then she abruptly turned off into the woods overlooking Sinking Valley, poults streaming after her in disordered array. She showed such a lack of fear that I began to wonder if she had seen me, but since I had been in clear view and since turkeys have excellent eyesight, I knew she had.

Several minutes after she disappeared down the mountain

slope, I heard a lone poult peeping on the power line behind me. I glimpsed it in the underbrush, but it fled further into cover as soon as it saw me. Thinking that the hen might return if I was gone, I walked on, but I could still hear the piercing peeps of the abandoned poult a couple hundred feet away. I only hope that the hen heard them too.

JUNE 5. It isn't often that we can verify a new mammal species on the mountain, but during the last two days we have had two sightings of fox squirrels. The first occurred as I was walking along Laurel Ridge Trail yesterday. I paused when I spotted a large squirrel trotting down the trail toward me. Closer and closer it came, moving slowly and deliberately, concentrating on where it was going and not on what was on the trail in front of it. At first I thought it was a gray squirrel, but as it neared me, I realized that it was larger than the usual gray squirrel and that it had a red border around its head and tail as well as an underside of red on its large fluffy tail. It paused within six feet of me, sat up on its haunches, studied me carefully, and then ran slowly and quietly off into the woods.

To make certain of my identification, I carefully wrote down all its field marks and behavioral characteristics which I checked against my mammal books back home. Only fox squirrels have a red underside to their tails, the books told me. They also prefer the ground to the trees and are altogether a larger, less excitable squirrel than the gray. I had been hearing for years that there were fox squirrels on our mountain, and a couple of times I thought I might have seen one. But it was never close enough to verify. Bruce and Steve had twice this spring reported seeing what they thought was a fox squirrel down in the hollow, but being a "doubting Thomas" I refused to believe until I saw it with my own eyes.

Today, as if to make certain that I would doubt no longer, Bruce and I saw not one but three fox squirrels leisurely chasing down the hollow road and through the woods as we drove up from town late in the morning. Again we observed the red undersides to their large tails and noted how they moved more deliberately than gray squirrels and stayed on the ground.

Whether they were eastern or western fox squirrels is another question since neither has been verified for our section of the county. But not long ago a valley neighbor told me that he and his brother had released several western fox squirrels, which they had caught in Ohio, on our mountain ten years ago. Also, the western fox squirrel has been extending its range in western Pennsylvania, so I'm inclined to believe that we have seen *Sciurus niger vulpinus*. On the other hand, the range maps show *Sciurus niger rufiventer*, the eastern fox squirrel, in the southern tip of our county, closer to us than the western fox squirrel.

Since the subspecies are difficult to distinguish, I remain content with the knowledge that we have a small population of fox squirrels on our mountain—whether western or eastern is immaterial except to biologists who have classified the eastern fox squirrel as "undetermined" in the Pennsylvania Biological Survey's report, *Species of Special Concern in Pennsylvania*.

JUNE 6. On this beautiful June day I took a female friend for a walk on the mountain. Along Laurel Ridge we had an excellent view of a solitary vireo with feathers in its mouth, my first verification of nest-building here by that species.

When we reached the top of the Laurel Ridge Trail where it joins both the First Field Trail and the Far Field Road, I scanned First Field as I usually do and spotted the head and neck of a hen turkey poking out of the tall grass far below like a periscope. "Do you want to try to see her poults?" I asked my friend. Although it meant walking a considerable distance downhill away from the direction we were heading, she was game.

Slowly we moved down the faint field trail that Bruce had

mowed last summer. I kept my eyes trained on the spot where I had seen the head and, just as we reached the spot, the hen erupted out of the grass at my feet while the poults fled off into the brush. One obligingly crouched for a moment at my friend's feet, providing her first view of a young turkey in her seventy years of life, before it ran off into the grass after its siblings.

Later, as if on call, the hen grouse in the old dump area started whining close to us while her chicks flew up and then off into the woods. I felt proud at how well the wildlife of the mountain had performed, even though my friend had initially been disappointed by the difficulty she had had in spotting songbirds now that the canopy has closed in. She did, however, add two new birds to her life list, first a pileated woodpecker as she was driving up the hollow, and then a field sparrow in the spruce trees at the top of First Field. Both are common species on the mountain and proof that what seems ordinary to one person can be extraordinary to another. Living in a small town, my friend is familiar with certain species of birds, such as mockingbirds, that are rare here; but she had never seen, until today, the deep-woods-loving pileated woodpecker or the common, but difficult to identify, field sparrow.

Despite her love of the outdoors, my friend is a gregarious person who likes to be going and doing most of the time. Just before she left, late in the morning, she took a final look around and asked hesitantly, "Don't you ever get tired of this place?" She meant, I think, that she would be lonely despite the obvious abundance of wildlife here. Most people do need the stimulus of human society, and by her comments my friend made me realize how different I must seem to others, even to those who appreciate the outdoors. But most people do not have the inner resources to live an isolated life—as I do by choice. Even those who fight hardest for wilderness areas come mostly from the suburbs and cities—to them the woods are a nice place to visit but they certainly wouldn't want to *live* there.

JUNE 7. Now that the last of the ephemeral woodland wildflowers has faded, the wildflower pageant has shifted to the fields. The predominant species are no longer native, such as the trailing arbutus and Canada mayflower, but instead are immigrants from European fields and cottage gardens.

Dame's rocket is the showiest of the late May and early June wildflowers, ranging in color from white to pale orchid to dark purple. It thrives throughout our grape tangle, on the slope beside the guesthouse, and along the edge of First Field. *Hesperis matronalis,* its scientific name, means mother of the evening, so-called by Pliny because its fragrance increases at night. This flower was cultivated by Roman matrons, hence the "dame's" part of its common name, and was brought by the Romans to England where it became a popular garden flower.

I have bouquets of dame's rocket in my living room, but I derive my greatest pleasure from watching it outdoors. The four-petaled, flat blossoms that form a showy cluster along the two-to-three-foot stems attract many species of colorful butterflies and moths. Today I kept a record of the number of species I saw feeding in the patch. First I watched a perfect tiger swallowtail butterfly unroll its long proboscis and push it down into the middle of each blossom to collect nectar. Then came two spicebush swallowtails followed by a red admiral, an American painted lady, a sphinx moth, and a cloudy-wing skipper. But the most common butterfly, by far, was the silver-spotted skipper. At any one time I could see several clinging to the flowers.

The buttercups and hawkweeds have also opened their golden blossoms in the fields. Neither flower is a friend of farmers because they can crowd out pasture grass and legumes. In addition, buttercups contain acrid juices in their stems and leaves which inflame the mouths and intestines of cattle and can spoil the flavor of milk and butter if too much is cut and dried with hay. Nevertheless, buttercups remain popular with children and poets and have many common names such as "king cups," "gold cups," "cuckoo flowers," "blister-flowers" and "butter-flowers." And as the botanist Claire Shaver Haughton wrote in her excellent and informative book about foreign wildflowers, *Green Immigrants,* "a field deep in gleaming buttercups delights the eye."

A field deep in golden or orange hawkweeds can also be beautiful. Their genus name, *Hieracium,* means "hawk"; people used to believe that hawks ate the leaves to improve their already keen eyesight. The species growing in our fields is called "king devil" *(Hieracium pratense).* Its stems and bracts are covered with bristly blackish hairs, and its leaves, which grow at the base of the stem, are fuzzy with white hairs. Another name for hawkweed is "Grimm the Collier," referring to the coal miners in England. Probably the black hairs reminded people of the grimy appearance of miners when they emerged aboveground.

No wildflower has had more folklore built up around it than yarrow, still another June-blooming field flower. Its fame stretches back to the siege of Troy. Achilles, so the ancients believed, used the leaves of yarrow to heal his wounded soldiers, thus its generic name *Achillea.* Its species name—*mille-folium*—is Latin for "thousand leaves."

Yarrow's fame spread among medieval herbalists who claimed that its leaves reduced toothache, induced nosebleeds, dispelled melancholy, controlled falling hair, coagulated blood, and lowered fevers. Contemporary scientists say there is no evidence for any of those beliefs, but they have discovered that if yarrow is planted in gardens, it acts as a pest repellent and an agent against blight. It even improves the flavor of vegetables grown nearby.

Although it was once thought that all wild yarrow came

from Europe, botanists have recently discovered that a native species, *A. lanulose,* is far more common than the European species. Only by using microscopic techniques, though, that reveal the chromosomes, can trained botanists tell the two apart. Nevertheless, whether the wildflowers of June are native or immigrants, they all have their own unique characteristics that add to the beauty of our fields.

JUNE 8. Grouse, grouse wherever I go on the mountain. First I was stopped, driving up the road, by a mother grouse with her chicks parading across in front of me. Then, along First Field Trail, I saw probably the same hen and chicks that I saw with my friend the other day. The hen's cries were so pitiful, as she tried to decoy me away from her fleeing young, that they would have wrung the heart of any mother. She scrunched down her body and ran around in circles like the proverbial "chicken with its head off." Once I was well past her, she climbed up on a limb, clearly visible to me, and started her "wuk, wuk" call to let her chicks know where she was.

My third sighting of the day was along the Short Circuit Trail, where hen and chicks both scattered; and the fourth was in a fern bed beside the Far Field Road. The hen ran off to the right of the trail; the six chicks that I counted to the left. I sat down and waited expectantly for her to call them back, but when she finally flew across the road, landed in the woods, and started calling, the chicks never made a peep. So I continued on my walk knowing that once I disappeared, mother and children would be quickly reunited.

JUNE 9. I have never seen the white woodchuck again, and I have not seen a fox since May 24. But at 10:30 on this windy, humid morning, a small red fox, still with prominent black guard hairs, walked across Laurel Ridge Trail thirty feet in front of me. It had its eyes on the trail and never looked up.

Then it plunged into the underbrush, and I could hear it moving off across the mountainside. No doubt it was one of the Far Field youngsters out on its own hunting expedition.

JUNE 10. Many people ask me if we have any pets on our farm and my answer, during the last couple of years, has been no. The assorted Nimrods who roam our mountain have shot them all. "But what kind of a farm is it without dogs and cats?" is the usual retort. "A naturalist's laboratory rather than a farm," I reply.

And so it is. Now that we have finished with domestic pets, the wild animals have come closer and closer to our house. Unfortunately, there are disadvantages to this—an incredible woodchuck population explosion and an abundance of deer and rabbits which forces us to choose between wild animals and a vegetable garden.

We have had to resign ourselves to the natural look in landscaping, growing only those flowers, herbs, vegetables, and shrubs that the wild animals don't like. This means that we have very nice onions, radishes, garlic, irises, daffodils, forsythia, lilacs, ferns, and violets. The rest of our "yard" is green grass, assorted weeds, and wildflowers. It would never win a suburban beauty contest. In fact, we would probably be cited for letting our lawn grow too high and for not keeping our flower beds neat and trim. Luckily we have no neighbors to complain and those who visit us are mostly nature lovers like ourselves, so they are as thrilled as we are when the animals come in close.

Our laissez-faire approach makes yard care supremely easy. The time we would ordinarily spend nursing along specimen plantings we can spend instead in happy contemplation of the beauty around us. Others can slave to keep their lawns fertilized, their flower beds and vegetable gardens weeded, and their ornamental trees and bushes pruned; we sit back and let nature take over. And believe me, nature needs little encouragement. After one rainy spell not long ago, we had to beat our way through grass that seemed to grow a foot in one day, searching for the asparagus bed behind the garage.

But that tall grass hides other treasures besides asparagus—fawns and young rabbits. Today Steve was trying to make some headway in cutting the foot-high grass between the guesthouse and the shed when he narrowly missed running down a small cottontail. He scooped it up and brought it in to show me. Its ears were laid back, its heart was beating wildly, and it screeched occasionally in fear of the big creature that held it. Eventually it calmed down a bit, enough to sit in the grass and pose in front of Bruce's camera before hopping off into the tall, sheltering grass.

Larger rabbits have been coming out at five o'clock every evening to feed, and when Bruce drives up from work, there is a scattering of bunnies all along the edge of the driveway. We usually eat our dinner watching them eat theirs—the tender tops of newly mown grass and weeds in our back yard. They pause to scrub their faces with their front paws and scratch their sides with their back paws. If we call out "hello," they barely pause to glance up and acknowledge us before bending back down to supper.

But loveliest of all is the trustful doe who has been bringing her yearlings to the flat, grassy area below our sitting room window ever since last fall. Earlier this month Steve glanced out his bedroom window and noticed the doe in a rather peculiar position, her head turned around and down. After carefully watching for a few minutes, he hurried into the kitchen to tell me that she had a newborn fawn with her. We could scarcely believe our eyes as we stood at the dining room bow window and watched her licking the small creature all over with her tongue while the fawn nursed.

"Real live Bambi in Walt Disney color," Steve commented. She continued licking and the fawn nursing for ten minutes. Then she moved off a few feet to begin eating the grass. The fawn wobbled close behind her, bent its small head, and also nibbled the grass. After a few more relaxed minutes of grazing, mother and child wandered on out of our range of vision toward the power line right-of-way. The doe had occasionally glanced up at us watching from the window, but never once had she appeared to be nervous about our presence. It was as if she were proudly showing us her newest offspring, or so it seemed to our wondering eyes.

For such scenes we will gladly sacrifice the right to own domestic pets.

JUNE 11. Early June is mountain laurel time here. Laurel Ridge Trail is rimmed by thousands of laurel bushes, and when they bloom as spectacularly as they have this year, enveloping me in a cloud of pink and white, I consider the sight the finest natural display our mountain produces. Every shrub, from less than a foot high to over ten feet tall, is covered with so many blossoms that they obscure the laurel leaves.

The loveliest example is the shrub with deep pink flowers I discovered hidden back in the woods several years ago. I found it again today, glowing like a flame amidst the white-blossomed shrubs surrounding it. I probably should report it to Richard A. Jaynes, author of *The Laurel Book,* who discussed the Connecticut Agricultural Experiment Station's breeding and genetic study of laurel species which led to the development of such new strains as Goodrich and Shooting Star. Both were discovered in the wild by mountain laurel enthusiasts who sent colored slides of them to Jaynes. There can be wide variations in the native mountain laurel, and it is those variants which are needed for the development of new strains.

Mountain laurel, *Kalmia latifolia,* is one of the seven native species of the genus *Kalmia* named after Peter Kalm, an eighteenth-century Swedish naturalist whose journal of his travels through Pennsylvania contained one of the first detailed accounts of mountain laurel. "Their beauty rivals that of most of the known trees in nature," he wrote.

The Swedish settlers in America called them "spoon trees" because the Indians made spoons and trowels of the hard, strong, but brittle wood of the mountain laurel. Other names are "poison laurel," "calico tree," "calico flower," "big-leaved ivy," and "ivy wood." In the southern Appalachians, where laurel sometimes grows thirty feet tall, its flowers are sold along the road by natives who call it "ivy" since southerners insist that the name "laurel" should be reserved for wild rhododendron.

"Poison laurel" is also an appropriate name because the leaves contain andromedotoxin, which causes salivation, weeping emesis with convulsions, and paralysis of the limbs in browsing domestic animals such as sheep and goats. Yet white-tailed deer often browse on it in late winter without any ill effects. This year I saw evidence that they did so here. Studies do show, though, that if deer are force-fed laurel leaves they will be poisoned too, but normally they do not eat enough to cause illness.

Mountain laurel has always been appreciated for its beauty, and it is the state flower of both Connecticut and Pennsylvania. Because it likes the acid, peatlike, clay soil of the Atlantic seaboard, it is only found in the wild as far west as central Ohio in the north and Louisiana in the south. Instead of transplanting it into our garden, we have followed the advice of the late naturalist Donald Culross Peattie: "Buy yourself an Appalachian hill, forget your gardening tools, your bales of peat, your soil injections of silicate of aluminum."

And in June our Appalachian hill wreathed in mountain laurel is lovelier than any formal garden.

JUNE 12. Green, that is what my world is, bright, brassy green, the kind of green that hurts your eyes when the sun shines. But since the sun hasn't shone for a couple days, the gray skies tone down the green.

Whenever there is a spell of rainy weather I tend to feel sorry for the creatures who can't retreat into the warm, dry walls of a house. Conversely, I feel sorry for myself, enclosed by the walls of a house. So early this morning, when it was cold (fifty degrees), damp, wet, and gloomy, I bundled up into my long-sleeved pullover, wool shirt, and hooded coat sweater and sat out on the front porch.

Our elevated front porch provides me with a ringside seat for the treetop wildlife behavior among the black walnut and black locusts which emerge from the lawn far below. Off to the left of the porch grows the huge old lilac bush that earlier attracted sphinx moths and courting hummingbirds and that now offers shelter to many bird species. A couple hundred feet down off the right side of the porch is the grape tangle and an old apple tree. Directly in front and below the porch are the stream, the driveway, the springhouse, the guesthouse, and two red maple trees in full leaf.

I brought out my coffee and a book to fortify me at seven, but one look into the trees and I forgot both. Despite the weather, life continued on with birds singing, working, courting, and feeding, and I was kept busy watching.

The robins were building a nest in the lilac bush and were not being particularly secretive about it. They didn't seem concerned by my watching them as they fluttered in noisily with long strands of newly mown grass hanging from their beaks. Then a phoebe landed on the locust tree, her beak crammed with insects for her brood in the springhouse nest. She didn't have to worry about the rain because her nestlings are well protected by the enclosed structure. I know that there are five little heads in that nest, with mouths gaping open, clamoring eagerly for food.

Eastern wood pewees, sounding like phoebes with a Southern drawl, called monotonously from the edge of the woods, a counterpoint to the continuous song of the red-eyed vireos. I located one silent bird of each species in the black locust trees, but they evidently preferred to do their singing in the woods.

A pair of northern orioles landed in a locust tree at 7:20, the first orioles of the day, and shortly afterward I heard the male singing. Later I listened to what I thought at first was their buzzing scold call, only it wasn't. Instead it was the preliminary to their brief mating, which I watched high up in the trees.

At 7:45 I heard the calling of the red-bellied woodpecker that haunts the Carolina poplar tree in the back yard. The call was so tantalizingly near, yet I could see nothing but leaves and branches. Then suddenly the woodpecker flew over to the same hole in the black walnut tree that the northern flickers occupied two weeks ago. And I had a good look at its red crown and black-and-white striped back as it poked its head in and out of the hole and looked around for perhaps three minutes. Then it flew back to a small, dead branch of the Carolina poplar where it called loudly and drummed not quite as loudly for a few minutes before bouncing off across the barn roof.

Next a gray squirrel slid quietly along a black locust trunk. It moved as if it knew I was there and didn't want me to see its aerial leaps, five feet across and thirty feet up from the ground, maneuvering over space and between trees until it reached the fourth locust tree.

From the ground the trunks are at least thirty feet apart, but as I watched the squirrel leaping from branch to branch, I became part of the branches laced against the sky. I imagined what it must be like to have a squirrel's agility, to know intimately each branch—how far out it can go before it must take its courage in hand and leap the distance to the safe section of the next branch.

There I sat, earthbound, rooted to a chaise longue on a

solid porch, and wished to have the airiness of a bird, the nimbleness of a squirrel. And then I heard a snort from the edge of the woods. I turned to look and glimpsed only the disappearing tail of a deer. I realized with a shock that it too is earthbound, but with such grace and agility as it slips through the undergrowth or leaps away from danger that it is closer to the squirrel than to me. It and all its kind can survive the gloomy weather without four dry walls while I searched the sky hopefully for a glimmer of light that would call me to the trails again.

But by 8:45 it was still gray and green out. My nose and my feet were cold. The walls were calling me in to wash the dishes and clean the house. I sighed and left the robins to build their nest, the orioles to court, the chorus of birds to continue and, most of all, the squirrel to leap, unfettered by fear and four dry walls.

JUNE 13. We tend to take the small, black-and-white, downy woodpeckers for granted because they are year-round residents that can be easily watched at our bird feeder during the winter. Yet until this spring, I had never found one of their nests. Then, in early June, I began to notice that the small, slippery elm sapling outside our kitchen door was constantly fluttering with movement from a pair of downy woodpeckers. They seemed to spend most of their time clinging by their feet to the leaves and pecking at them—strange behavior for birds that usually glean food from tree trunks. Upon closer examination, I discovered many elm leaves had become the rolled-up homes of insect larvae. The downies had decided to reap the bonanza, but they were not eating it themselves. Instead they would pack their bills full of larvae and fly off into the woods. Obviously they had a nest there.

Several days later I started up the Guesthouse Trail on my usual early morning walk and almost immediately heard the unmistakable cries of downy woodpecker nestlings. After a quick look around, I spotted the twenty-foot-high stub of a dead tree. Near the top was a woodpecker hole. I settled myself at the base of a chestnut oak tree thirty feet away from the stub, and within minutes a male downy woodpecker landed at the edge of the hole and poked his head in. Shortly thereafter the female appeared and did the same thing. Each time a parent arrived, the incessant chippering of the nestlings increased in volume.

I watched the nest for more than an hour and then, on June 9, I observed it from 7 until 8:15 A.M. That day I could see the heads of the nestlings as they took the food their parents brought in. The female seemed to be doing most of the feeding. Although the male would land frequently in front of the hole, he would have nothing to do with the young birds. Such tantalizing action by the male was apparently designed to turn the nestlings into fledglings.

Today the tree stub was deserted. Again it was my ears that put me on the alert. I heard a loud calling and then noticed a movement on a double red maple tree thirty feet from the nest tree. A downy woodpecker fledgling was clinging to the right tree trunk about six feet from the ground.

This time I sat down at the base of the nest tree to watch. The female parent landed next to the fledgling and began feeding it. For more than an hour I watched. I saw only that one fledgling being fed by the female. There was no sign of either the male or the other fledglings. Yet, according to all the bird books I consulted later, fledged downy woodpeckers are strong fliers that stay in the nest area for a week and are fed by both parents. The books also indicated that a male fledgling usually has a red patch on top of its head. The one I watched did not.

So the fledgling was probably a weaker female who had been the last to leave the nest. Although she could cling to the tree trunk, she could not fly well, and so the female parent had been left to look after her. Possibly the male was caring for the stronger fledglings nearby, but if so I never saw them. As usual my wildlife observations had left me with more questions than answers, which is why I continue to seek, expecting that along the way I will sometimes find.

JUNE 14. Last night David called us out at dusk to see a box turtle on the lower path across the power line right-of-way. She was digging a hole to lay her eggs in by bracing herself with her front feet while scooping out the dirt with her hind feet. The hole was perfectly round, not much larger than a quarter, and a couple of inches deep. Although most box turtles dig their nests and lay their eggs between 6:00 and 9:00 P.M., this female was clearly behind schedule. It was already 9:30 and she had barely begun her excavation in the light sandy soil, so we left her to it.

Usually box turtles dig their holes and lay their eggs outside their home ranges after searching for just the right soil. Years ago Mark and I discovered the remains of box turtle eggshells at the base of a power pole on top of Laurel Ridge. That female had obviously liked the mixture of fine stones and sand piled around the pole. The eggs had lain underneath the ground until September when the youngsters had hatched, emerged from the soil, and hidden themselves from predators until they were fully developed, in the meantime defending themselves by emitting a bad odor. Very few naturalists have ever found the wily young box turtles. Lucille Stickel, over her long study period, discovered over two hundred adults but only twenty-six young.

Today I returned to the site, having noted its exact location. To my surprise, it was easy to find because it was prominently marked by a hill of loose earth. Underneath, I assumed, were four to five thin-shelled, flexible eggs shaped like oblong ping-pong balls. Evidently she had not tamped down the earth firmly with her shell as box turtles usually do, so I helped her along by tamping it myself. To my eyes, at least, it looked far less obvious although it remained a bare spot of earth in the center of a grassy path, and I very much feared that any sharp-eyed predator such as a skunk or a raccoon would easily find the eggs. But we shall see.

JUNE 15. I never thought that I would watch nature finish a sequence I last observed in late January. But that is what happened this morning.

I was halfway up the Guesthouse Trail when I suddenly noticed what appeared to be a platoon of chirring gray squirrels racing up and down nearby trees and making daring leaps from the topmost branches. At first I stood, then I sat, and finally I lay on my back watching the treetop antics, trying to decide if once again gray squirrels were courting. According to wildlife biologists, gray squirrel females often have two litters a year in our area.

Although they moved so fast it was difficult at first to sort them out, after a while the field narrowed to what was probably one female and a male who fended off the occasional forays of the other five contenders. She, on the other hand, always climbed out to the thinnest branch and faced him down when he approached her by chirring loudly and threatening to nip him. He would chir just as loudly and also make threatening motions. But for over an hour she eluded his advances by leaping off the branch at the last moment and landing on another tree, streaking down the trunk, and stirring up the waiting males in the underbrush below.

Once she led her entourage to a tree five feet from where I was sitting, and all of them raced up the tree after glancing over at me. Apparently sex was more important than safety. As they chased and leaped from branch to branch, pieces of bark rained down around me. Finally a male cornered her again, drove off all his rivals, and mated with her high in the rocking branches of the tree. After about ten seconds she shook him off, streaked down the tree, and the chase resumed.

Again she was isolated by a male, whether the same as the

first or another I could not tell, and they faced off, chirring loudly. Twice the male approached and retreated. But at last he mated with her, this time on a lower tree branch.

For the third time she escaped, ran down the tree, and set the other males in pursuit. After a short scuffle in a laurel bush, which I could not see clearly, the noise subsided and the squirrels disappeared.

Unless I am mistaken, I witnessed the conclusion of the preliminary courtship rites I had observed on a cold winter morning almost six months before in another place by different squirrels.

JUNE 16. On this humid, warm day two events of note took place. The first occurred this morning when a fledgling flew up from the fern bed beside the Far Field Trail. It had long pinkish legs, a streaked white-and-brown breast, and a brown back and head. It landed and froze in a small red maple sapling, which gave me ample time to study it closely from every angle. At first I thought it might be an ovenbird, but then a male towhee came in "chewinking" at a great rate.

I moved in still closer to examine the fledgling's tail and noticed its resemblance in miniature to the tail of an adult male towhee—black with white edges. My moving prompted the adult to fly above me and then to fly into the sapling below the fledgling which began chirping softly. Finally it fluttered back down into the ferns as the father continued "chewinking" until I walked away.

The second event occurred at dusk when I spotted the first fireflies of the year. This dark, moonless night heavy with humidity was ideal for fireflies since they don't like to compete with moonlight.

Fireflies have intrigued humanity since ancient times. The Chinese and Japanese believed they emerged from decaying grasses, and Aristotle thought they developed from worms on peas. Throughout the Middle Ages there was great interest in fireflies, continuing into modern times—particularly after the discovery of the New World, where they were often spectacular. Jamaica, for instance, is famous for its diversity of luminous night-flyers, with more than fifty native firefly species.

Western travelers who explored the tropical regions of India, Southeast Asia, the Philippines, and New Guinea were awed at the sight of trees along the rivers alight with fireflies blinking on and off in perfect synchrony. In 1727 Sir Hans Sloane observed this phenomenon along the Meinam River near Bangkok, Thailand. Howard Ensign Evans, in his excellent book *Life on a Little-Known Planet,* quotes the American biologist Hugh M. Smith, who wrote: "Imagine a tree 35 to 40 feet high thickly covered with small ovate leaves, apparently with a firefly on every leaf and all fireflies flashing in perfect unison at the rate of about three times in two seconds, the tree being in complete darkness between flashes."

Today researchers who study synchronous fireflies hypothesize that because of the heavy vegetation in the tropics the male fireflies congregate in the trees so that they can be seen more easily by the searching females. After all, sex *is* the reason why firefly species flash throughout the world.

In the eastern United States most fireflies are winged, and the flashing lights are produced by both sexes. Usually the males flash in the air while the females answer from the ground or on a perch when they are ready for mating. Once they have mated they lay luminous eggs in the grass or soil which hatch into luminous larvae (commonly called glow-worms) in thirteen to forty days. The larval stage can last two years or longer and is followed by a pupal stage. Finally, the pupal skins split to emit fireflies that live only a few months.

Fireflies are not flies but soft-bodied beetles of the family *Lampyridae.* They produce their light by the interaction of a substance called luciferin with oxygen and an enzyme, luciferase. Molecules of adenosine triphosphate, commonly called ATP, which is a high energy compound found in all living cells, are a catalyst for the flashes. More than two hundred species use light for mating, but the color and rate of flashing vary from species to species.

The common smaller fireflies of the eastern United States belong to various species of the genus *Photinus* and are almost identical in outward appearance. According to Chuck Fergus, James Lloyd, a biologist at the University of Florida, has discovered that they actually emit slightly different light signals at slightly different distances from the ground in order to attract only the females of their own species. Here on our mountain the largest and most common firefly is *Photuris pennsylvanica*, although there are several of the smaller genus, *Photinus*, as well.

Years ago, when our sons were young, collecting fireflies and putting them in a jar to wink away the dark hours was a favorite pastime. To them, fireflies were a natural part of life. Then one summer we had visitors from the West Coast. "What are all those blinking lights outside?" the six-year-old girl asked her mother. And her mother, who had been reared in the East, suddenly remembered her own childhood.

"Why, fireflies," she answered, and tried to explain to her little daughter what they were. But her eight-year-old son did not wait to hear. He grabbed a jar from one of our boys and dashed off into the darkness. To me, the wonder of children and fireflies have remained inextricably linked ever since.

JUNE 17. This year Mark is devoting his time to compiling a biological inventory of our property which he calls BIO-PLUM, short for "A Biological List of Plummer's Hollow." He has divided it into six sections: amphibians, reptiles, mammals, birds, ferns, and woody plants. Today he enlisted my help for the fern section since years ago we had spent time inventorying those species growing down in the hollow. At that time I had identified nine species there, including the New York, cinnamon, interrupted, Christmas, lady, sensitive, and both the spinulose and marginal wood ferns, all of which are common.

The ninth and least common was the rattlesnake fern which grew in only one small location about a quarter mile up the road, near a trickle running off our neighbor's side of the mountain. Because of the imminent threat of lumbering in that area I was thrilled, earlier this month, to find a brand new location for rattlesnake ferns on our side of the road much further up the hollow and far more abundant. A succulent fern of the family *Ophioglossaceae,* this particular representative of the grape fern genus has a stemmed sporophyll or fertile spike protruding high above its single triangular leaf.

The other common species here grow on the drier slopes—bracken and hay-scented ferns. The latter has spread across our power line right-of-way and emits a sweet smell to the area, especially shortly after it opens in May.

But this evening Mark was eager to show me two new ferns he had found. At the top corner of First Field he pointed out one ebony spleenwort growing in an infertile, dry area where the weeds and grasses were short. I was amazed that it had grown there for so many years without my noticing it. Once we began to look more closely, we found several more plants in the same kind of dry habitat in other areas of the field.

I was even more chagrined at Mark's second find—common polypody—growing on the bank side of the road across from the corral. So many times I have walked slowly past that area, scanning for new plants. How could I have missed those small, dainty ferns all these years? Again, the more we looked, the more we found. After such an experience I am certain there must be more than the thirteen fern species we have so far discovered here since there are 100 species in the northeast and midland states and 10,000 species throughout the world. Nevertheless, *Pteridophyta,* which includes ferns and their allies, is still the smallest phyllum of the plant kingdom, representing the first true vascular plants with stems, roots, and leaves but without flowers and seeds.

The reproduction of ferns was long a mystery to ancient peoples who believed they were magical plants. Today the reproductive processes of the various fern species are understood, although each species differs in its approach. The marginal wood fern, for instance, produces approximately 52 million spores in one plant. The thousands of fruit dots found on the undersides of fertile marginal fern leaflets contain numerous spore cases, each with sixty-four spores. On a dry, windy day when the spores are ripe, the fruit dots burst and catapult the spores out onto the ground. Those that find a bit of shady moisture at the proper temperature develop, cell by cell, into heart-shaped gametophytes over a two-week period. Each gametophyte contains both male and female organs, the male organs near the apex or pointed end, the female organs near the notched end. Those organs unite, the female organs bending down toward the male organs, when the weather is wet. The fertilized egg develops, anchored to the bottom of the female organ. It then grows a root down into the soil and a stem up through the gametophyte's notch. It is the stem that produces the first tiny leaf, and with the development of an independent plant, the life cycle of the fern is completed.

One Mother's Day the boys transplanted several species of newly unfurled ferns in our back garden, and all have thrived. Ferns are not fussy like woodland wildflowers; they often grow beautifully in a shady location. There is nothing so cooling to the senses on a warm day as the sight of green ferns growing along a wet marsh or stream—one of the loveliest essences of an Appalachian woods.

JUNE 18. Back on May 21 I noticed an ovenbird flying down in the brush beside our Guesthouse Trail with strands of dried grass in her bill. I say "her," because the female ovenbird selects the nest site and builds it by herself while the lookalike male guards the area and gives an alarm call whenever an intruder appears.

I was excited about the possibility of finding an ovenbird nest because it is one of the most unusually constructed of songbird nests. It is built on the ground of natural materials that blend in perfectly with its surroundings, and it has a rounded roof, much like a Dutch oven, which is how the ovenbird got its name.

I made no attempt that day to look for a nest. Obviously the mother was still building it. If she is disturbed, an ovenbird will sometimes desert a partially built nest or even her eggs in the early stages of incubation, so I noted down exactly where I had seen her and resolved to search another day. In the meantime, I learned that ovenbirds frequently build their nests beside woodland trails much like the Guesthouse Trail. Sure enough, three days later I located the nest beside the trail in a few minutes, probably because it already had one egg in it. That egg was white, speckled at one end with brown, and the white stood out against the brown of the nest. A couple of bracken ferns grew near the entrance of the "oven," which was neatly roofed with leaves and grass constructed in an arch over the branch of a small mountain laurel shrub.

The following day, when I checked the nest, it had three eggs, one of which was robin's egg blue. I knew it could not be a cowbird's egg, even though cowbirds often parasitize the ovenbird, because their eggs are a beige-gray color. Despite the alien egg, the mother continued to lay her usual cluster of five eggs, and she was sitting on them and the mysterious blue one by May 29. However, on June 10, only the five ovenbird eggs hatched, so I never was able to figure out the ownership of the blue egg. It may have belonged to a black-billed cuckoo because they, like the European cuckoo, sometimes lay their eggs in other birds' nests.

I visited the nest several times during the next eight days.

Usually the mother was brooding the nestlings, sitting with her brown side parallel to the nest opening, just as she had done when incubating the eggs. Even though I knew its exact location, it was almost impossible to see the nest with the mother in place. When I brought visitors up for a look, I had to point directly at it with a long stick before they could distinguish it from its surroundings. Of course, the mother never moved, and I never touched her or her nestlings because I did not want my scent to attract predators. As a matter of fact, in my almost daily morning check of the nest, I saw the nestlings alone only twice.

But today I found the nest empty. An adult ovenbird sat on a tree branch across the trail from the nest scolding me. No doubt the little ovenbirds were hiding on the ground nearby, since many of them do leave the nest after eight days, enticed by their parents' offering food just out of reach. Slowly they hop out, one by one, and each parent takes half the brood to finish raising.

The young fledglings walk along the ground with their parent, who provides food for itself and its offspring by turning over leaves on the forest floor in search of such delicacies as snails, slugs, earthworms, beetles, crickets, ants—in fact any insects that live there. While insects make up most of an ovenbird's diet, a very small proportion (slightly more than one-fiftieth) consists of seeds and small fruits. Until the fledglings are three weeks old, the parent bird continues to feed them; after that they feed themselves, with some help from their parent. At five weeks of age they are on their own. Not only do the parents abandon them, but they leave the natal woods altogether, since ovenbirds have only one brood a year.

Next spring that same pair of adult ovenbirds may again nest beside our Guesthouse Trail, because both male and female like to return to their old breeding grounds, according to Harry Wilbur Hann. First the males arrive and establish territory, then the females return. It is the territory that attracts them, however, not the resident male. Monogamous relationships occur only by chance, depending on whether last year's male arrives first and is aggressive enough to drive out its rivals, and on whether last year's female gets to the area ahead of all others of her sex. During the years Hann studied and banded ovenbirds, two pairs did re-mate, he reported.

So, in all likelihood, next year's parents will not be the same. But whether or not the pair I observed returns next spring, I know that our woods will continue to be a haven for those brown-backed warblers with long pinkish legs and orange crowns.

JUNE 19. Today I added another mammal to Mark's BIO-PLUM, only it was a body, not a live creature. On the Laurel Ridge Trail lay a dead long-tailed weasel. It had been recently killed because we found dried blood on its one side. Other than that, though, its body was still a lustrous dark brown with a yellowish white breast. No wonder its fur is highly sought after.

Long-tailed weasels are much maligned because of their alleged ruthlessness and fondness for killing. On the contrary, they merely cache the extra prey they kill for another day. But many people I have talked to around here tell me they kill every one they see, which may be why it has taken me seventeen years to find one. In all the years we kept a flock of free-ranging chickens, not once was our less than secure hen house raided by a weasel. Instead, we had marauding raccoons which ate the twenty-four Muscovy ducks that roosted in our barn one summer, and then, the following year, invaded our chicken house and ate the entire flock. Raccoons, as we discovered, are far more destructive to domestic fowl than long-tailed weasels, yet they continue to be popular with the local people, probably because of their endearing looks as juveniles and their grownup role as prey to predator men who hunt them with dogs on moonlit autumn nights.

Long-tailed weasels, on the other hand, are not poster-cute

and do not provide excitement for baying dogs and hunters. Furthermore, they have the audacity to compete with man as a fierce predator on both domestic and game animals such as cottontail rabbits and gray squirrels. Most of their diet, though, is nonthreatening to humanity's interests—meadow voles, white-footed mice, chipmunks, shrews, moles, birds, reptiles, and insects. They, in turn, are killed, but not always eaten, by numerous predators including several owl and hawk species, red and gray foxes, coyotes, bobcats, domestic cats, and the larger snake species. Because of their strong, musky odor, weasels do not make enjoyable eating and so are often abandoned as today's specimen was. So which animal is the wanton killer, the long-tailed weasel or the unknown assassin who killed, but did not eat the weasel?

JUNE 20. Seduced by the setting sun and lilting birdsong, Bruce and I went for a walk after dinner. We hiked along the edge of First Field, talking quietly, and then we swung left into the woods, intent on following the Far Field Trail.

Loud rustling noises in the laurel understory brought us to a standstill. They did not sound as if they were being made by deer. "Turkey?" Bruce whispered, remembering how we had listened to a large flock of scratching turkeys in the same area one spring evening.

"Skunk?" I suggested. Just that morning I had watched a foraging skunk at the top of Laurel Ridge Trail and had been impressed by the amount of noise one small animal could make.

Both of us were wrong. Suddenly an enormous black bear lumbered out on the trail one hundred feet ahead of us. He turned to face us, down on all four feet, sniffed a couple times in our direction, and then loped back into the woods, "woofing" loudly. Scarcely able to contain our excitement, we stood listening until the bear sound faded away. That bear had been big—three hundred to four hundred pounds, Bruce estimated—and I was glad I had not been alone as I was last spring when I saw a medium-sized black bear on Laurel Ridge Trail.

We continued toward the Far Field still moving quietly so we could appreciate the evening chorus of wood thrushes. Just as we neared a curve in the trail, we heard a loud "woof" about twenty-five feet below us in the woods to our left. It was difficult to decide which of the three of us was more startled. Once again we had inadvertently prevented the bear from crossing the trail.

As we peered down at him, he lumbered off a short distance. But even after he stopped moving, he continued "woofing" like a startled deer. He also seemed to think we could not see him as long as he remained still. Despite his enormous size, his actions reminded me of a small child playing hide-and-seek.

Finally we moved and he did too, still "woofing" as he headed due east while we went south. He was probably the same bear I had encountered on June 20 last spring, ambling along the power line right-of-way. He had grown considerably since then, and I assume he was looking for a mate, as mature black bears do in June.

After seeing tonight's enormous specimen I can believe the wildlife biologists who claim that our state has the biggest black bears in North America.

JUNE 21. Gradually June has leveled off into the somnolence and quietude of summer. On this official last day of spring it is eighty degrees by seven in the morning. Already the fresh look of spring is gone. The mountain laurel blossoms are fading, and there are large holes in almost every tree leaf from the chewing of gypsy moth caterpillars. The mountain looks dusty and dry because there has been no dew in the grass for several days. Heat covers us like a stuffy blanket that we cannot escape. Even the birds are quieting down.

But at the first gray light of dawn, a single field sparrow sang its spiraling song and then was joined by the undistinguished droning of countless chipping sparrows. As Bruce would say, "The dicky birds rule the mountain by sheer numbers." I then heard a robin and a wood thrush before falling back to sleep until 6:00 A.M.

The ephemerals in the woods, including the long-blooming rue anemones, are gone, along with the precious ephemerality of the season itself. How can it be over so quickly? How many more springs will I see? Every year those questions become more urgent to me as I near the half century mark. But I will never have enough of spring no matter how long I live, and I hope that if there is a heaven, it is an eastern spring in North America. Surely not even God could improve on that.

Spring is my idea of paradise on earth which no vision of heaven could exceed. When I hear people speak of their reward in heaven after their sojourn on earth, I am impatient. Enjoy this earth and especially this spring, I think. It may be all of heaven we will ever know.

But spring had one more surprise for me. As I entered the Far Field this morning, I spotted two adult turkey heads above the grass quite close to me. Then the air ahead of me erupted with young turkeys flying up in every direction as the hens dashed off in a flutter. They continued to take off in groups and to fly into trees in the field and along its periphery while I tried to count the bewildering numbers. Thirty-five seemed like a reasonable estimate—no doubt the result of three or four families joined together.

I remained standing, feeling as if I had had a proper ending to my spring, when one hen landed over on the far side of the field and emitted calls that sounded like a begging puppy. Twenty-four poults sailed out of the locust grove toward her, and all were finally lost from my sight, but not my ears, by the grass.

Our mountain becomes wilder year by year as the wildlife numbers increase. Conversely, civilization laps ever closer to the mountain's edges—louder and louder, more and more insistent—and the mountain serves as an island of refuge for the wild things driven from the fields and housing developments of the valley.

When we first moved here, the number of wild creatures was far smaller and the silence far greater. All we ever heard were the trains in the valley. Then the town bypass road at the base of our mountain was opened, and the steady drone of trucks joined the train whistles. Lately a large corporation bought up a nearby quarry that is consuming a large portion of a neighboring mountainside. The quarry owner only respects Sundays. Otherwise the machinery screeches night and day, and we complain about the excessive noise pollution that most of our fellow easterners know more about than we do.

We can never escape the commercial jetliners crisscrossing our sky space thirty thousand feet above, nor the helicopters that beat down close, searching for illicit marijuana plants or inspecting the electric lines. Then there are the small military jets that skim so low that the treetops bend, practicing their war games over what they perceive to be an underpopulated area.

The message is clear. No matter how isolated our home may seem to most people, all that we do here is open to the prying eyes of aircraft overhead, and the peace we crave is shattered by burgeoning civilization and its noise. We and the animals, crowded onto a mountain island that shrinks from year to year, are more closely acquainted than we were seventeen years ago. The rewards of these close encounters partly compensate for our loss of peace and privacy, but we sometimes long for quieter climes. No doubt the animals do too.

I awoke near 11:57 P.M. when summer officially arrived, but I heard no bacchanalian rites such as our Celtic ancestors might have performed. Midsummer's eve was always the highlight of primitive people's lives, so closely entwined were they with the cycle of the seasons. To most of us, seasonal changes mean little more than a switch from central heating to central cooling as we go about our human-centered lives, forgetting, until drought or flood or hurricane hits, that it is, in the end, nature and not the self-important ways of humanity that will control our destiny.

Selected Bibliography
Index

Selected Bibliography

General

Beck, William M., Jr. "Suggested Method for Reporting Biotic Data." *Sewage and Industrial Wastes* 27 (1955): 1193–97.

Borland, Hal. *Hal Borland's Twelve Moons of the Year*. New York: Alfred A. Knopf, 1979.

Burroughs, John. *Far and Near*. Boston: Houghton Mifflin, 1904.

Brooks, Maurice. *The Appalachians*. Boston: Houghton Mifflin, 1965.

Genoways, Hugh H., and Fred J. Brenner, eds. *Species of Special Concern in Pennsylvania*. Pittsburgh: Carnegie Museum of Natural History, 1985.

Morgan, Ann Haven. *Field Book of Animals in Winter*. New York: G. P. Putnam's & Son, 1939.

Olson, Sigurd F. *Listening Point*. New York: Alfred A. Knopf, 1958.

Van Dyke, Henry. *Fisherman's Luck and Some Other Uncertain Things*. New York: Scribner's & Son, 1899.

Birds

Bent, Arthur Cleveland. *Life Histories of North American Birds of Prey*. 2 vols. New York: Dover Publications, 1961.

———. *Life Histories of North American Flycatchers, Larks, Swallows, and Their Allies*. New York: Dover Publications, 1963.

———. *Life Histories of North American Nuthatches, Wrens, Thrashers, and Their Allies*. New York: Dover Publications, 1964.

———. *Life Histories of North American Thrushes, Kinglets, and Their Allies*. New York: Dover Publications, 1964.

———. *Life Histories of North American Wagtails, Shrikes, Vireos, and Their Allies*. New York: Dover Publications, 1965.

———. *Life Histories of North American Woodpeckers*. New York: Dover Publications, 1964.

———. *Life Histories of North American Wood Warblers*. 2 vols. New York: Dover Publications, 1963.

Brown, Leslie. *Eagles, Hawks, and Falcons of the World*. New York: McGraw Hill, 1968.

Gabrielson, Ira Noel. "A Study of the Home Life of the Brown Thrasher." *Wilson Bulletin* 24 (1912): 65–94.

Hann, Harry Wilbur. "Life History of the Ovenbird in Southern Michigan." *Wilson Bulletin* 49 (Sept. 1937): 145–237.

Heinrich, Bernd. *One Man's Owl.* Princeton, N.J.: Princeton University Press, 1987.

Kroodsma, Donald E. "The Spice of Bird Song." *The Living Bird Quarterly* 5 (Spring 1986): 12–16.

McAtee, Waldo Lee. "Woodpeckers in Relation to Trees and Wood Products." *U.S. Department of Agriculture and Biological Survey Bulletin* 39, 1911.

Nice, Margaret Morse. *The Watcher at the Nest.* New York: Macmillan, 1939.

Peterson, Roger Tory. *A Field Guide to the Birds.* Boston: Houghton Mifflin, 1980.

Stokes, Donald W., and Lillian Q. Stokes. *A Guide to Bird Behavior: In the Wild and at Your Feeder.* Volume 2. Boston: Little Brown, 1983.

Terres, John K. *The Audubon Society Encyclopedia of North American Birds.* New York: Alfred A. Knopf, 1982.

Walker, Lewis Wayne. *The Book of Owls.* New York: Alfred A. Knopf, 1974.

Wink, Judy; Stanley Senner; and Laurie Goodrich. "Food Habits of Great Horned Owls." Hawk Mountain Sanctuary Association press release, 1987.

Zeleny, Lawrence. *The Bluebird: How You Can Help Its Fight for Survival.* Bloomington: Indiana University Press, 1976.

Insects, Spiders, and Worms

Borror, Donald J., and Richard E. White. *A Field Guide to the Insects of America North of Mexico.* Boston: Houghton Mifflin, 1970.

Darwin, Charles. *Formation of Vegetable Mould, Through the Action of Worms, with Observations of Their Habits.* New York: D. Appleton, 1882.

Dethier, Vincent G. *The World of the Tent-Makers: A Natural History of the Eastern Tent Caterpillar.* Amherst: University of Massachusetts Press, 1980.

Emerton, James H. *The Common Spiders of the United States.* New York: Dover Publications, 1961.

Evans, Howard Ensign. *Life on a Little-Known Planet.* New York: E. P. Dutton, 1968.

Fergus, Chuck. "Thornapples." *Pennsylvania Game News* 60 (August 1989): 51–54.

Holland, W. J. *The Moth Book: A Guide to the Moths of North America.* New York: Dover Publications, 1968.

Hutchins, Ross E. *Insects.* Englewood Cliffs, N.J.: Prentice-Hall, 1966.

Klots, Alexander B. *A Field Guide to the Butterflies.* Boston: Houghton Mifflin, 1951.

Klots, Elsie B. *The New Field Book of Freshwater Life.* New York: G. P. Putnam's Sons, 1966.

McManus, Michael L., and Roger T. Zerillo. *The Gypsy Moth: An Illustrated Biography.* USDA Home and Garden Bulletin no. 225. Washington, D.C., 1978.

Nichols, James O. *The Gypsy Moth.* Harrisburg: Pennsylvania Bureau of Forestry, 1980.

Preston-Mafham, Rod, and Ken Preston-Mafham. *Spiders of the World.* New York: Facts on File, 1984.

Scott, Jack Denton. "The Wonder of the Worm." *National Wildlife* 6 (Aug.–Sept. 1968): 33–35.

Stokes, Donald W. *A Guide to Observing Insect Lives.* Boston: Little, Brown, 1983.

Williams, Ted. "Ardis and the Gypsies." *Horticulture* 57 (Sept. 1979): 19–26.

Mammals

Aleksiuk, Michael, and Antony P. Stewart. "Food Intake, Weight Changes, and Activities of Confined Striped Skunks (Mephitis mephitis) in Winter." *American Midland Naturalist* 98 (Oct. 1977): 331–41.

Doutt, J. Kenneth; Caroline A. Heppenstall; and John E. Guilday. *Mammals of Pennsylvania.* Harrisburg: The Pennsylvania Game Commission, 1967.

Ferris, Chris. *The Darkness Is Light Enough: The Field Journal of a Night Naturalist.* New York: Ecco Press, 1986.

Godin, Alfred J. *Wild Mammals of New England.* Baltimore, Md.: Johns Hopkins University Press, 1977.

Hamilton, William J., Jr. "Winter Activity of the Skunk." *Ecology* 18 (April 1937): 326–28.

Hamilton, William J., Jr., and John O. Whitaker, Jr. *Mammals of the Eastern United States.* Ithaca, N.Y.: Cornell University Press, 1979.

Henry, J. David. *Red Fox, the Catlike Canine.* Washington, D.C.: Smithsonian Institution Press, 1986.

Jackson, Donald Dale. "It Is About Time the Shrew Stood Up to Be Recognized." *Smithsonian* 16 (Oct. 1985): 147–53.

Kelker, George Hills. "Insect Food of Skunks." *Journal of Mammalogy* 18 (1937): 164–70.

Macdonald, David. *Running with the Fox*. New York: Facts on File, 1987.

Merritt, Joseph F. *Guide to the Mammals of Pennsylvania*. Pittsburgh: University of Pittsburgh Press, 1987.

Rood, Ronald. *How Do You Spank a Porcupine?* New York: Trident Press, 1969.

Wishner, Lawrence. *Eastern Chipmunks: Secrets of Their Solitary Lives*. Washington, D.C.: Smithsonian Institution Press, 1982.

Woods, Charles A. "Erethizon dorsatum." *Mammalian Species,* no. 29 (June 13, 1973): 1–6.

Plants

Bailey, L. H. *How Plants Get Their Names*. New York: Dover Publications, 1963.

Bierzychudek, Paulette. "Jack and Jill in the Pulpit." *Natural History* 91 (March 1982): 22–27.

Cobb, Boughton. *A Field Guide to the Ferns*. Boston: Houghton Mifflin, 1956.

Cook, Robert E. "Fragile Blossoms of Spring Aren't Shrinking Violets." *Smithsonian* 8 (March 1978): 64–71.

Correll, Donovan Stewart. *Native Orchids of North America North of Mexico*. Waltham, Mass.: Chronica Botanica, 1950.

Gibbons, Euell. *Stalking the Wild Asparagus*. New York: David McKay, 1962.

Grimm, William Carey. *The Shrubs of Pennsylvania*. Harrisburg: Stackpole and Heck, 1952.

————. *The Trees of Pennsylvania*. Harrisburg: Stackpole and Heck, 1950.

Harned, Joseph Edward. *Wildflowers of the Alleghanies*. Oakland, Md.: n.p., 1936.

Haughton, Claire Shaver. *Green Immigrants*. New York: Harcourt Brace Jovanovich, 1978.

Hylton, William H., ed. *The Rodale Herb Book*. Emmaus, Pa.: Rodale Press, 1974.

Jaynes, Richard A. *The Laurel Book: Rediscovery of the North American Laurels*. New York: Hafner Press, 1975.

Knutson, Roger M. "Plants in Heat." *Natural History* 88 (March 1979): 42–47.

Lust, John, ed. *The Herb Book*. New York: Bantam Books, 1974.

Meeuse, Bastiaan J. D. "The Voodoo Lily." *Scientific American* 215 (July 1966): 80–88.

Peattie, Donald Culross. *A Natural History of Trees of Eastern and Central North America.* 2d ed. New York: Bonanza Books, 1966.

Peterson, Roger Tory, and Margaret McKenny. *A Field Guide to Wildflowers of Northeastern and Northcentral North America.* Boston: Houghton Mifflin, 1968.

Rickett, Harold William. *Wildflowers of the Northeastern States.* 2 vols. New York: McGraw Hill, 1966.

Reptiles and Amphibians

Babcock, Harold L. *Turtles of the Northeastern United States.* New York: Dover Publications, 1971.

Bellis, Edward D. "The Influence of Humidity on Wood Frog Activity." *American Midland Naturalist* 68 (1): 139–46.

Conant, Roger. *A Field Guide to Reptiles and Amphibians.* Boston: Houghton Mifflin, 1958.

Dickerson, Mary C. *The Frog Book.* New York: Dover Publications, 1969.

Pope, Clifford H. *Turtles of the United States and Canada.* New York: Alfred A. Knopf, 1939.

Stickel, Lucille F. "Populations and Home Range Relationships of the Box Turtle, Terrapene C. Carolina (Linnaeus)." *Ecological Monographs* 20 (October 1950): 350–73.

Wassersug, Richard. "Why Tadpoles Love Fast Food." *Natural History* 93 (April 1984): 60–69.

Index